Building Information Modeling

FOR DUMMIES®
A Wiley Brand

by Stefan Mordue,
Paul Swaddle, and David Philp

Building Information Modeling For Dummies®

Published by: **John Wiley & Sons, Ltd.,** The Atrium, Southern Gate, Chichester, www.wiley.com

This edition first published 2016

© 2016 John Wiley & Sons, Ltd, Chichester, West Sussex.

Registered office

John Wiley & Sons Ltd, The Atrium, Southern Gate, Chichester, West Sussex, PO19 8SQ, United Kingdom

For general information on our other products and services, please contact our Customer Care Department within the U.S. at 877-762-2974, outside the U.S. at (001) 317-572-3993, or fax 317-572-4002. For technical support, please visit www.wiley.com/techsupport.

For technical support, please visit www.wiley.com/techsupport.

A catalogue record for this book is available from the British Library.

Library of Congress Control Number: 2015950110

ISBN 978-1-119-06005-5 (hardback/paperback) ISBN 978-1-119-06007-9 (ebk)

ISBN 978-1-119-06008-6 (ebk)

10 9 8 7 6 5 4 3 2 1

Contents at a Glance

Table of Contents

Introduction

*P*erhaps you keep hearing about Building Information Modeling (BIM)
and want to work out what all the fuss is about. You may be a complete beginner in digital design and construction, looking for the basics. You may be a confident CAD user who wants to understand what BIM brings to the party. Or maybe you're already very experienced in implementing BIM processes and you're just interested in what we have to say in *Building Information Modeling (BIM) For Dummies.* We understand that you could be at various stages of knowledge and levels of experience. BIM is a process and it needs people like you to fuel it.

BIM isn't just a buzzword. It's actually been around for a long time, but the foundations to make it really work weren't in place before now. We've seen BIM generate new efficiencies and new challenges too, and we've also seen how construction is hit hard by global economic recession. Our aim is to give you the push you need to start, accelerate, or maximize your use of BIM. It's our great pleasure to guide you through your BIM implementation journey.

About This Book

More than likely, you've encountered a lot of conflicting, overly complicated, and unhelpful content about BIM online and in print. We intend to make *Building Information Modeling (BIM) For Dummies* different, cutting through all the noise and providing you with the clear advice and practical guidance that you need to make BIM a success in your job, whatever your level may be. Think of this book as a reference guide that addresses just what you need to know about BIM.

For the purposes of this book, we want to make it clear that we use the terms *model* and *BIM* to mean quite different things. Throughout the book we refer to BIM only as the concept of information modeling for buildings. We use terms like *3D CAD, 3D model,* and *geometric/geometry model* interchangeably.

We structure this book to start with the basic concepts and ideas. We then gradually introduce more complicated, detailed, or supplementary content. Within this book, you may note that some web addresses break across two lines of text. If you're reading this book in print and want to visit one of these sites, simply key in the web address exactly as it's noted in the text, pretending that the line break doesn't exist. If you're reading it as an e-book, then you can just click the web address and the page will open in your default Internet browser.

One thing we should note is that we're all based in the UK, and that's where our expertise comes from, but we've tried wherever we can to balance this with a US perspective throughout and a global reference where relevant.

BIM is such a fast-moving target that during the writing process things shifted and new documents were released. We focus on pointing out the most current versions of guidelines and protocols that we reference at the time of writing.

Foolish Assumptions

Now, in order to write a practical guide to BIM and not just another boring textbook, we make some assumptions about you, our dear reader. Those assumptions are as follows:

- ✔ You have some knowledge of the construction industry, probably in a professional capacity, in design, contracting, or surveying.

- ✔ You're a student, a client, or the owner of a building from a totally different industry, but you want to have a better grasp of a typical construction project, including the contractual relationships and expected information exchanges.

- ✔ You work for a company that has an interest in incorporating BIM into your processes and workflows.

You don't need to know how to use CAD software in order to gain insight from this book, because we think the software platforms are only one part of BIM and that model technicians are just one role in many BIM responsibilities.

You also don't need to have any project management background. We explain the majority of management concepts throughout the book, but you may find understanding how to implement BIM processes easier if you've run jobs or managed a project team.

Icons Used in This Book

We include icons in the left-hand margins that highlight particular parts of the text you may want to remember or pay close attention to. Those icons include the following:

This icon is practical, giving you extra information about how to do something or how to save time in larger tasks.

Remember icons highlight snippets of the book that you should commit to memory. Think of them like those bright sticky notes you may have on your desk.

Try to always read text with a warning icon. With this icon, we point out some pitfalls so that you can avoid BIM disasters, and we also emphasize actions you should take to prevent running into problems.

We flag the more technical information with this icon so that you know which parts are extra to the core content. You don't have to read these bits to put the rest of the book into practice, although you may find this information interesting.

This icon points out supplemental information online at www.dummies.com/extras/bim.

Beyond the Book

With your purchase of *Building Information Modeling (BIM) For Dummies* in print or e-book form, you have access to more exclusive information online. From great checklists on BIM processes to quick practical articles, you can find so many helpful pointers at www.dummies.com/extras/bim.

In addition, every *For Dummies* book includes a Cheat Sheet with handy information that you may want to consult on a regular basis. You can access the Cheat Sheet at www.dummies.com/cheatsheet/bim.

Where to Go from Here

Every *For Dummies* is modular, which means that you don't have to read it in order from cover to cover. If you're new to the world of BIM, we recommend that you start with Part I because it provides lots of the getting-started information that forms the foundations of BIM implementation. If you have a basic foundation of BIM, you can focus on Chapters 5, 6, or 7.

If you have the time, we suggest that you do read this book from cover to cover to get a complete overview of BIM and the reality of its implementation. You can see the overall picture when you're able to finally step back and benefit from the wealth of specific knowledge in this book.

If you're confident that you know all about the basics, you can jump into any part by going directly to it. For example, if your area of interest is the BIM mandates and protocols around the world, you can head straight over to Chapter 9. Alternatively, if you're looking into the legal aspects of BIM, you can flip to Chapter 14. If you're not sure where to start, consult the table of contents or index for a topic that interests you and then start reading.

Part I

Getting Started with Building Information Modeling

Go to www.dummies.com for bonus information about BIM and most any other topic that interests you.

In this part . . .

- Find out how to explain what BIM is in a really simple way and understand what you really need for BIM implementation.

- Appreciate that BIM isn't just for buildings, but suitable for all kinds of infrastructure projects too, and look at examples of how more efficient processes are impacting the industry.

- Make it easy to interrogate your project data by filling the model full of structured information, which other project team members can use for a variety of different uses and applications.

- Use the right modeling tools to develop accurate 3D object information and see the benefits of detailed modeling.

- Set up a common understanding of what BIM is for and agree on the fundamentals of BIM with your colleagues and project teams.

Chapter 1

Defining Building Information Modeling (BIM)

*T*he construction industry has been doing things the same way for thousands of years. Concrete is poured and set, bricks are stacked on top of bricks, and systems for heating and water are designed around corners and over multiple floors. For way too long, the construction industry has done a lot of these processes in isolation. At its worst, the construction industry brings some people involved in the construction of an asset like a building or a bridge onto the project just in time for their part, and the project team has to work around decisions or redo work, often on-site and under pressure of project deadlines.

Even in some of the most collaborative schemes, communication between different teams still has a long way to go, and the other users of building data and outputs, like clients and facility managers, are sometimes the last to know. The quality and quantity of data they receive on a project can vary wildly. What you need is a way to involve the entire project team earlier and coordinate all the project information in clear and accessible forms.

If only a combination of processes and technology existed that provided the framework to improve communication and data exchange across the construction industry, no matter how large or complicated projects may be. Well, interestingly enough, you're in luck. This chapter serves as your jumping-off point to that very process: Building Information Modeling, commonly shortened to BIM.

Explaining BIM in Plain Terms

Here we provide a good definition for the term BIM so that the three members of your author team and you are on the same page. Frustratingly, BIM actually has lots of definitions, many generated by various organizations, because the subject has changed over the years. To prevent any confusion, we present you with our own definition that we think really clearly explains what BIM is and what it isn't.

Most people agree that the acronym BIM stands for Building Information Modeling, but a few folks argue for Building Information Management (and, to be honest, some other alternatives too). (The next section takes a closer look at what the three letters in BIM mean.) More often than not, though, BIM is now an accepted acronym, so you don't need to break it down further anyway, just like RAM for random access memory. We think that BIM is a process, so we could easily use both Modeling and Management in our definition. Here it is:

> BIM is a process for combining information and technology to create a digital representation of a project that integrates data from many sources and evolves in parallel with the real project across its entire timeline, including design, construction, and in-use operational information.

Examining the A-B-Cs of BIM

BIM stands alone as a word in its own right, and you can feel confident using it, instead of having to say "Building Information Modeling" in full every time. But when it comes to understanding what BIM really is and explaining it to other people, those three letters can be a very useful place to begin. The following list gives a bit more detail about the A-B-Cs of BIM, or, more accurately, the B-I-Ms!

B is for banana

One of the best ways we've found to describe BIM to someone without any knowledge of it is to grab a piece of fruit. Explain that you could develop a perfect 3D replica of the fruit in digital modeling programs or even by 3D-printing a copy, but that's only one kind of representation of the fruit. It doesn't include any of the fruit's *data*; for example, its sugar content, calories, use-by date, country of origin, whether it has Fairtrade certification, and so on. The 3D object on its own isn't enough to represent the fruit. It isn't Banana Information Modeling. That's the difference between 3D CAD and true information modeling.

- **B:** Because the B in BIM stands for *building,* think of this as the verb *to build,* and not just the noun, as if BIM was for just physical, discrete buildings. In fact, you can apply BIM to infrastructure, civil engineering, and landscape, along with large-scale public and private projects.

  You're modeling a process, the act of building something. Refer to Chapter 2 for more information on what the B in BIM means and for help on BIM for infrastructure.

- **I:** The I in BIM is about understanding that unless you have *information* embedded throughout the project content, the work you're producing is telling only half of the story.

  You don't even really need to worry about the modeling in order to start applying BIM; you can put the processes and data exchanges into practice long before drawing work begins on a project. The real value in BIM is the ability to interrogate the model and find the data you need, when you need it. Turn to Chapter 3 for some great examples of information modeling from other industries, like aeronautics and automotive racing.

- **M:** The M stands for *modeling.* This aspect of BIM probably has the most history, and hundreds of programs for representing the built environment using 3D CAD techniques and virtual design and construction (VDC) are available. (In fact, the majority of free resources on BIM, especially in the United States, can put too much focus on the 3D modeling aspects of BIM.) Chapter 4 is about how the visual model should evolve in detail, but only as much as you require for the relevant output. The model should allow the output of whatever plan/section or perspective or walkthrough or 3D-printed model that you require.

One of the simplest ways you can explain BIM is that the project should be *built twice:* once fully modelled digitally and then again for real on the construction site.

After you comprehend the definition of BIM, the next step is to grasp what BIM is actually trying to achieve. BIM processes aim to make you (and the construction industry as a whole) more efficient, and to allow project teams to make savings in terms of cost, time, and carbon, and removing waste across the timeline. Chapter 5 provides a really simple overview of what BIM is trying to do and some of the key fundamentals you need to know.

Understanding the Requirements for BIM

Here's a list of what you really need for BIM implementation to thrive:

- **Digitization:** You have to be confident that the future of the industry is digital. Think about how technology has evolved in most industries and how in your experience of the construction industry you may have noticed that it's still traditional and paper-based. BIM implementation requires a change of direction, toward new tools and software and a digital future.

✔ **The right foundations:** In order to build advanced BIM processes, you need the firm bedrock of efficient systems for communication, information exchange, and data transfer. Think about what practical changes you may require and even the type of projects you focus on. We show you how you can describe your BIM readiness in terms of levels of maturity. Chapter 7 discusses the importance of having a foundation before you implement BIM.

✔ **Process:** What's wrong with what you're already doing? We hope you can see where you can make improvements in your current processes, and that moving toward BIM implementation should have a positive effect on your business. Some essential elements to collaboration exist, and in Chapter 8 we show you an example of best-practice work flow and an explanation of some of the key acronyms you'll encounter.

✔ **Technology:** You need to ensure that you have the right technology to support your BIM aims and objectives. Technology includes software and hardware. Having the right technology enables you to work in a digital environment. In Chapter 21, we show you different types of BIM platforms and software, with some important examples and discussion points for when you have to make decisions.

✔ **Training:** All the processes, frameworks, and documents in the world won't help if people don't understand them and can't use the tools and methods you're implementing. So a key requirement is to support all the technology and protocols with dedicated and personal training. We point you towards some great resources you can look to for help in Chapter 22.

✔ **Incentives and business drivers:** *Incentives* are what motivates and encourages you and your organization to undertake BIM, whereas your *business drivers* refer to processes that are vital for the continued success and growth of your business. Some business drivers may be outside business drivers; for example, economic conditions that a company can't always influence. The UK is mandating BIM from 2016, and the United States, although still behind, demonstrates huge potential for standardization. In Chapter 9, you can read about the UK Government Construction Strategy and where the BIM mandates came from, and compare it with BIM uptake in the United States and across the rest of the globe.

✔ **Standardization:** For BIM to thrive, you need interoperability. Interoperability is a term that's important in BIM-speak. *Interoperability* is ensuring that you can use the outputs someone else in the project team has produced, because you're all using standard formats. Other BIM standards exist, along with a range of recommended protocols, guidelines, and specifications for the properties of objects you use in your models in the form of information exchanges. In Chapter 9, we help you navigate through these documents and show you how everything could evolve.

A very brief history of BIM

Even though there's been a recent push to implement BIM and a realization of its benefits, BIM isn't a new concept. The earliest use of the term *building modeling* was in the 1980s, in a paper that predicted that model objects would connect to relational databases full of different kinds of information. And long before that, college research teams were developing computer modeling techniques with buildings in mind. Even just on the graphical side, university research has had a significant impact on modeling advances.

Software companies have been developing tools for built environment professionals to design, plan, render, and analyze buildings and structures for decades. Although most have focused on 3D geometric modeling systems,

the largest platforms have been exploring how to make the most of data science and the properties of building products too. The first use of the term *BIM* to describe all this goes back as far as the 1990s. The awareness, investment, and supporting documentation have all increased dramatically in the past few years, though.

For more information on the history and theories of BIM, we suggest that you check out some great books by the fathers and godfathers of BIM such as *The BIM Handbook* by Chuck Eastman, Paul Teicholz, Rafael Sacks, and Kathleen Liston (John Wiley & Sons, Inc.) and *Building Information Modeling: BIM in Current and Future Practice* by Karen Kensek and Douglas Noble (John Wiley & Sons, Inc.).

Considering BIM Plans and Strategies

Having a clear plan and strategy is essential to the success or failure of your BIM journey. You'll need an overall strategy for encouraging BIM in your office or on-site. Use the BIM protocols and frameworks to refine and improve your processes and quality assurance, and develop individual BIM execution plans for particular projects.

So that BIM processes have the best possible chance of becoming everyday practice, you want to make a start with your current team and your next project. In Chapter 13, we show you what having a BIM strategy really means and what benefits you can expect from new methods of working. To help you do this, we also present a couple examples of different BIM strategies:

- **BIM in the UK:** You can use the UK's suite of BIM documents in combination with your preferred tools and supporting platforms to achieve BIM Level 2 and what it's going to take to progress to Level 3.

- **BIM in the United States:** In the United States the same clarity of a national approach doesn't exist, but we direct you to a number of useful protocols and guidelines from certain states and BIM organizations, so that you can begin to build an efficient set of BIM processes and workflows.

Like everything in life, BIM also has some associated risks that you need to be able to identify. Some of those risks include

- **Digital security:** Sensitive information about the operation of assets
- **Intellectual property misuse:** Answering who owns the property
- **Risk and liability:** Recognizing who is responsible if something goes wrong

Chapter 14 discusses these challenges, what you can do to avoid them, and how to handle them quickly if you should encounter them.

Measuring the Real-World Benefits of BIM

Say that you've won over some key decision makers in your organization and they need you to produce a business case for BIM. As part of your business case, you need to justify the *capital outlay,* which relates to the money your organization spends to implement BIM. You also must consider upheaval that will come from new technology, new team structures, and even new staff. Not only that, but you probably have to demonstrate return on investment (ROI) as quickly as possible.

Your boss is going to want to know how much BIM is going to cost. BIM needs to generate savings and efficiencies that make it worthwhile. In Chapter 15, we pass on some solid examples of BIM benefits that aren't just aims for the future but exist in the real world today, including the following:

- **Better information:** Because you're going to be working with digital data and methodologies in the office or on the job site, the accuracy and currency of your information will improve, including precise quantity take-off and the ability to set the site out such as the asset's position, levels, and alignment from the model.

  Not only that, digital information also allows you to test and validate the data far more quickly than with traditional processes. As the model evolves, instant awareness of the impact of changes at any point in the project leads to better assessment and rapid decision-making.

- **Data exchange across the project timeline:** BIM can help you to avoid data loss over the course of a project. At many points of information exchange, you can use project data more collaboratively with little waste or duplicated effort.

What's even more important is that multiple roles and disciplines can use the same data on the project, including cooperative working with the supply chain and project participants further down the timeline, like facilities management and operations teams.

✔ **Communication:** BIM is your best chance to give your clients the built asset that they actually want and to output the deliverables that meet their own objectives, from slick visualizations to high-quality carbon data. Through a combination of 3D and nongraphic data, you can understand more about the built environment than ever before. Even better, you can also test out ideas in the safety of the model.

✔ **New efficiency:** The potential accuracy of BIM and the chance to refine engineering long before ground is broken on-site means that projects can begin to exploit new concepts like off-site manufacturing (OSM) where manufactures can deliver pre-built construction elements to site.

✔ **Carbon saving:** You can calculate statements about energy use and embodied carbon with new levels of detail. By running simulations and testing lifecycle concepts in the model environment as early as design and pre-construction stages, you can be more confident about the future performance of your asset. You can also have greater certainty over the project program and the likely issues that could arise.

✔ **Health and safety:** By improving information at the front end of BIM (including getting contractors and subcontractors on-board early), you can understand areas of risk in the project, especially where dangerous activities will take place, and achieve high levels of safety during later phases of the project. Throughout this book you can find examples of construction and site delivery of BIM, not just office-based BIM for designers. Chapter 16 specifically looks at some of BIM's impact on construction, especially the potential for BIM to improve health and safety on-site.

As well as including all the information about BIM's effect on projects today, we take a good opportunity to understand the future of the industry and where new technology like augmented reality (AR) could take BIM and digital construction in Chapters 18 to 20.

The construction industry is finally being disrupted by innovation and new business methods. It won't be long before the buildings and projects you're working on are more connected than ever. You may have heard the term *smart cities,* and BIM is one of the main generators of the embedded digital information required to achieve the connected globe. Through the addition of more smart building sensors, and what's called the Internet of Things, your understanding of how people really use the built environment (and your own projects) will improve beyond anything you could have previously imagined.

Encouraging BIM in Your Workplace

The amount of software and industry documentation you throw at an office doesn't matter, because so much of BIM implementation is about changing real-world processes and engaging individuals, with their various concerns, agendas, and opinions. How do you go about integrating BIM into real teams with real people?

People are the pulse of BIM, and you need to understand that the same BIM and the outputs it can generate are going to be used by different (and new) roles in the industry, at different times and in very different ways.

You can encourage people to embrace BIM by

- **Leading by example:** Be a BIM champion and lead by example with your commitment and enthusiasm.
- **Showing support:** Give support and encouragement by identifying and providing any training needs.
- **Communicating:** Deploy simple but clear messages about why and how you're implementing BIM.
- **Providing feedback:** Listen to other staff members and provide any reassurances that they may need around fear of change and the unknown.

There are various processes to BIM and many potential users involved. In more detail, Chapter 12 looks at encouraging BIM processes, and Chapter 17 focuses on BIM users and roles from inception to demolition (and beyond).

Our experience with BIM

We co-authored this book because we all have varied experience, from large contractors to small architectural practices. We've seen everything from the genesis of BIM for landscape to the development of ground-breaking software and documentation, including the various protocols and standards that the industry needs to communicate better, to evolve new methods, and to increase innovation.

BIM will continue to impact all areas of construction design, build, finance, management, and operation, and the relationships between all the parties involved in a project, both cooperatively and contractually. We think everyone can work together in more collaborative ways, toward a creative tomorrow that makes the most of diverse groups of people. Digital cooperation and access to information isn't just the heart of BIM; it's the heart of a connected, global society.

So are you ready? We want to make sure that you don't miss the BIM boat!

Chapter 2

Explaining the Building Part of BIM: It's Not Just Buildings

In This Chapter

▷ Introducing the "B" in BIM

▷ Recognizing the types of projects BIM is suitable for

▷ Exploring the use of BIM for infrastructure

▷ Delving into BIM as a process

*B*IM can seem like a bit of a strange term, and part of the reason it can be so difficult to explain what BIM means is that the letters don't always help you out. This chapter, Chapter 3, and Chapter 4 take each of the letters of BIM in turn and look at what they mean. This chapter focuses on the B in BIM.

As we discuss in Chapter 1, the B in BIM stands for *building,* which is true of most definitions of BIM. To avoid any misconceptions, this chapter makes sure the B in BIM doesn't restrict your view of what BIM is capable of.

Understanding What Building Means

What do you think of when you hear the word *building*? You may think of a physical building like an office, school, stadium, hospital, or house. In that case, BIM refers to information modeling for a single building, including all of the geometry and data for architectural and structural design, mechanical and electrical engineering, and so on.

Actually, building can mean a lot more than just that. The following sections explain that building is a misunderstood word and that BIM can actually be used in many varied industries and projects. If you think of building as a verb, not a noun, you can see that Building Information Modeling is a process, not just a final product.

Building isn't a helpful term

What makes understanding the building part of BIM difficult is that the word *building* isn't clear: it can mean different things to different people.

Try to describe what a building is. Doing so isn't easy. You can say that buildings are manmade structures, but what separates a building from a statue or monument? You can say they're permanent constructions with walls and roofs, but you'll be able to think of temporary buildings you've seen and also tunnels that have walls and roofs. In fact, one of the best ways to describe a building involves describing things that aren't buildings, and even that's confusing. Is a bridge a building? Is the Eiffel Tower a building?

The term *building* originally comes from ancient words for house. That's why people can think of buildings just as spaces they use for living or working or leisure.

Building as a verb, not a noun

If you think of building as a verb, meaning the same as construction or the process of putting things together, then that begins to expand what BIM can apply to. Then BIM isn't just suitable for buildings, it's the act of building things, such as the following:

- ✓ Bridges
- ✓ Railways
- ✓ Highways
- ✓ Utilities

You can also imagine how it's suitable for other built environment sectors like

- ✓ Land surveying
- ✓ Landscape architecture
- ✓ Tunneling
- ✓ Mining

Some people say that the B in BIM proves that it works only for buildings and just doesn't work for their discipline like tunneling or highways, but this simply isn't true. Every sector is at different stages of exploring BIM, and great examples already exist of those industries using BIM on live projects, so you shouldn't get hung up on the word *building*. For example, the Virtual

Construction for Roads (V-Con) project is a European initiative to improve data exchange across civil infrastructure teams by using BIM processes and it's changing road procurement in the Netherlands, Sweden, and France.

BIM also isn't just about architecture. Although the building design and construction industries have been the first to adopt the BIM processes and protocols as a group, BIM works for offshore projects, civil engineering, and infrastructure too. The documentation and support is increasing quickly for every sector.

Think of the B in BIM as meaning building as in the verb *to build,* the action of constructing things. Doing so helps you to understand BIM's reach in two ways.

- ✔ **It increases the sectors that information modeling applies to, sectors that build other things than just buildings.** BIM has been successfully demonstrated in

    - Architecture and building design

    - Civil and structural engineering

    - Energy and utilities

    - Highway and road engineering

    - Landscape and land surveying

    - Offshore and marine architecture

    - Rail and metro transport engineering

    - Services and engineering

    - Tunneling and subway architecture

    - Urban master-planning and smart city design

- ✔ **It demonstrates that building is a process.** It's not a one-time exchange of data; it's many exchanges over the life of a project. The majority of model inputs are going to be in the design and construction phases, and the majority of information outputs will be extracted during handover, use, and asset maintenance. The information modeling for the building process could start on day one and still be going strong years later.

In the same way that describing one building that sums up all the buildings in the world is really difficult, summing up BIM using just one example of how it's been applied is impossible. You can use BIM for every kind of construction project, from giant bridges to manmade islands and even rollercoasters! BIM is a term that has become popular gradually, but it could have just as easily been Construction Information Modeling or Project Information Modeling.

Thinking about the built environment

The *built environment* is very varied and broad in its scope and includes lots of structures that aren't buildings. When you're talking about BIM, make sure that you're not just talking about architecture and the architecture, engineering, and construction (AEC) industry. A lot of the diagrams and visualizations you see in BIM presentations are of shiny skyscrapers or complex building forms, but people are using BIM workflows elsewhere in the built environment in other ways:

- ✔ **Infrastructure:** *Infrastructure* is the network of systems that keep things moving, whether that's water, gas, electricity, traffic, or Internet data. The design, construction, and maintenance of these structures need to use the whole lifecycle approach of BIM. For instance, Crossrail (www.crossrail.co.uk) is the largest construction project in Europe and, among many projects, involves the tunneling of 26 miles of brand new underground subway lines. Every aspect of the project, from tunnel engineering to new underground station designs, has used innovative BIM processes for data management and lifecycle operation.

- ✔ **Geographic information systems:** Most built environment projects begin with site and land survey information. You can use *geographic information systems (GIS)* to visualize mapping and geolocational data so that the site information becomes part of the BIM. This is vital for city-scale projects. You can then add existing recorded data to the model, so that you can predict the impact of projects on traffic management, population density, or economic factors. It's really exciting when you start to think about doing it on a national or international scale.

- ✔ **Landscape architecture:** Landscape has been one of the most neglected sectors in terms of products and platforms that support the detail of landscape projects. Don't think that landscape is just tree and plant selection either; most landscape designers are involved in site modeling, level sculpting, and the overall aesthetics and performance of a scheme. For large infrastructure projects, the scale of forestry, wildlife, or water management can be epic. The platforms still leave a lot to be desired, but more landscape architects are able to coordinate their information with the rest of the construction team.

We should point out that our focus is primarily on Western design and construction practice, especially looking at BIM in the UK, Europe, and North America. However, a huge amount of BIM uptake occurs in Asia-Pacific regions, South America, and Africa. Increasingly, more information is becoming available about BIM implementation in those territories. We hope that the publisher of *BIM For Dummies* will let us do another edition to keep you updated!

Seeing How BIM Can Help You

Whatever type of project you're working on, you can apply the methods and processes of BIM to generate new efficiencies. Don't forget that you'll be building a digital representation of every aspect of the project. Some of the data is drawn, much of it in the form of information embedding.

Chapter 3 looks at information modeling and Chapter 4 at geometric, 3D-CAD modeling, and you may be thinking already that BIM sounds complicated, but you're familiar with a lot of the concepts already. This chapter demonstrates how making BIM processes second nature on your projects can benefit your work flow and the wider industry. The following sections look at some of the key incentives for using BIM processes and help you to make a decision about whether BIM is suitable for your project.

Don't worry about the various standards and protocols just yet. You may need time to understand all the detail and to digest some of the key documents and standards, especially the ones for your location, but everything will make sense eventually.

What does BIM do for building?

BIM can have numerous varied impacts on the work flow of a project. Here's a list of just a few of them:

Making design easier

The design phases of a project are one of the areas where the greatest reductions in wasted effort and rework can be made with BIM. From initial concept sketches based on client briefings to technical decision-making and product selection, design can be made easier.

Design efficiency increases through the use of pre-authored objects with embedded properties and relationships, including master template information for costs, carbon information, vendor manufacturing data, and performance specification values. Chapter 10 provides more detail on the development of BIM objects, and Chapter 17 takes a look at how the design team fits into a BIM project lifecycle.

Making coordination simpler

The digital building provides a single source of data, which simplifies managing all of the information, figures, and dimensions on a project. BIM makes it simple to coordinate drawn and nongraphical content. Chapter 10 describes this concept, including terms like *federated model,* which means that you can understand the impact of your design and construction decisions on everyone else involved in the project.

Ensuring construction is safer

One of the major drivers of BIM, in all applications but especially infrastructure adoption, is improving safety. This means site safety and awareness of potential issues, but also refers to making decisions as early as possible with health and safety in mind and to designing out risk and modeling safe construction and maintenance scenarios. Chapter 16 shows you how BIM collaboration can make a really positive difference in project health and safety.

Analyzing energy use

Busy buildings and modern construction infrastructure cost a lot to run, especially with rising gas and electricity prices. You can achieve one of the biggest cost savings for a built asset not by shaving off dollars from design fees or construction costs, but during use of the asset and its operational lifecycle. The largest cost is energy usage, so being able to model the carbon, thermal, and environmental strategy of a built asset and experiment with various options is hugely beneficial. Refer to Chapter 15 for more information.

Managing and maintaining the built asset

Another way to think of building the information model is that you need to embed all the information in the digital representation of the physical asset so that it can be managed and maintained in the long term. Focus on outcomes and what information you need later in the project's life. No longer will a building come with 40 boxes of paper drawings and spares; it should have a digital model embedded with data and clear procurement information. In Chapter 17, you can find out about making BIM work for facility management (FM).

Will BIM work for your projects?

You still may not be sure whether BIM is right for your infrastructure project. Perhaps your project isn't a traditional built asset, or it's something on a gigantic macro-scale, or else it's going to involve only a few teams working on just the early stage of concept design.

Implementing BIM work flows makes sense for all stages of a project and for all sectors and disciplines, even if you're not working on them all or with everyone else involved in a project. Even if the project lead isn't making a coordinated effort to use BIM processes and collate data, supply your information as if that was the case. You'll be doing everyone a favor and probably encouraging all participants to up their game.

You need to ask yourself if BIM is going to work for your projects and if the potential efficiencies or savings are worth the investment and related change that will be required. Figure out how you would approach the project using traditional methods and compare this to using only digital data and coordinated BIM tools and platforms. The answer may come down to the size of the project, but even the smallest constructions can be made more efficient.

For example, if you're developing a house extension, you may not implement full-scale BIM, but you can still improve your information exchanges with other team members. Don't forget that in some projects BIM usage may be an actual requirement for involvement in the team.

Here are some examples of building and civil projects where BIM is still hardly off the ground:

- ✔ **Projects at a scale thousands of miles in area:** You need to break these down into a significant number of smaller lengths for most software platforms to cope, which results in complex coordination. Managing the impact of change in one area on its surroundings can be very tricky.

- ✔ **Projects that are fundamentally two-dimensional:** Some examples include track layout and design on national rail networks. Many industrial manufacturers, fabricators, and suppliers have a long way to go before providing 3D object-based information.

- ✔ **Projects that involve the management of sites still in daily use:** Examples include rail, road, and airport maintenance improvements with traffic management requirements. The urgency and complexity of these projects can make front-end time savings the priority, not long-term lifecycle benefits.

Software platforms for BIM generally need to understand the concept of objects. Object-oriented technologies allow each piece of data to link with many other objects, in webs of connections known as *relational databases*.

Looking at Infrastructure and BIM

BIM adoption has been slower among infrastructure and civil engineering professionals. Here's a list of potential reasons:

- ✔ **No real incentive to share information:** The traditional "You do your bit, I'll do mine" is still very common in civil and infrastructure work. Specialist subject areas are technically very complex.

✔ **Perception that BIM may delay making site progress:** Whereas it *is* obviously cost effective to design out a clash between pipework and an architectural element in a building project before getting on-site, on a highways project engineers see getting around these issues as standard activity. This is especially true when projects are urgent or working to very tight timescales. Infrastructure clients sometimes see designing out all the clashes ahead of time as inefficient when in fact it could save millions of dollars in the bigger picture.

✔ **Lack of software designed to coordinate information at the scale of civil engineering projects:** Tools and platforms exist with infrastructure in mind, but they're generally targeted at one profession or discipline and for specific project sizes.

✔ **Information standards are rare and not enforced in the majority of projects:** A stalemate results, where teams just continue with existing processes and no overall management of the entire built asset exists. In simple terms, you need to ensure that the trains will fit in the stations.

You can use this train example to explain the need for communication and coordination. In 2014, France's national rail operator SNCF invested millions in new trains for regional travel. However, the survey of station dimensions was left incomplete, so the trains were designed too wide. This resulted in 1,300 stations needing to be "shaved" in order to fit the new, wider trains. The French government used the unfortunate story as an incentive to encourage information sharing across public-sector organizations and operators.

One of the objectives of the V-Con project for European roads that we mention earlier in this chapter is to produce a standard data-exchange structure for civil road engineering and future management. Having the standard structure in place hopefully will encourage software vendors to develop more advanced tools. The cultural change of users wanting to share their data will take longer to develop!

Developing Building Processes and BIM

BIM isn't just technology and it's definitely not just software. BIM is a best-practice process and therefore can impact project management and procurement just like some methodology, such as PRINCE2 or Agile, may completely restructure the delivery of a computing project.

Detailed discussions about BIM processes and your ability to explain them can easily become the focus. Bring the attention back to the outputs and outcomes of the project and the benefits of using BIM for asset information. The following sections begin to look at how you can develop BIM processes for the whole timeline of a built environment project. BIM applies right across the lifecycle and can improve design, construction, and operation.

Investigating design

BIM has the potential to make the design process easier and more efficient. Basic benefits of digital modeling include the ability to check if one designer's work clashes with another, being able to try lots of iterative designs out and fully understanding their impact, and better energy modeling and analysis.

- ✔ **Clash detection:** Projects that use BIM need the whole project team to work together, and each discipline will be developing a model in isolation. This can lead to coordination issues like overlapping systems or designs that can't be built because of other components getting in the way. When the models are brought together, it's important that problems are resolved and communicated through clash detection tools.

- ✔ **Multiple design options:** BIM lets you build digitally before you ever have to try things on-site. This opens up lots of possibilities, so you can test your ideas and work through many structural, engineering, and design concepts. The benefit of BIM is being able to interrogate these concepts in terms of their cost or complexity by using intelligent BIM objects. Then the entire project team can review the design at regular stages.

- ✔ **Energy analysis:** The construction industry is gradually becoming more sustainable, aiming to reduce energy use and waste on projects. BIM during design phases allows you to understand the impact of design decisions on energy use, overheating, and air circulation through energy analysis tools.

Using BIM in construction

BIM has many benefits during the construction phase of built projects. BIM can be used to schedule and plan out the construction process, including the movement of vehicles and plant machinery. The design decisions made in the model and increased precision of measurement should result in less wastage and higher accuracy during installation, along with the ability to explain difficult construction details.

Using the model as a communication method improves project teams' ability to collaborate and coordinate the work being done on-site. The model can also be used to calculate and manage the cost and time constraints of the project. In the long term, BIM will move toward automating the process of code approvals and building regulations too.

Operating the built environment

BIM can reduce costs during the operation phase of buildings, because the model forms an as-built record of all the systems constructed and installed. If the model is kept up to date, then BIM becomes an ongoing process to track maintenance, issues, and changes through the life of an asset. You can alert operational teams when systems are about to fail, pass their warranty date, or when they require maintenance or replacement according to a pre-written schedule.

As built assets become more automated and require more advanced building management systems, the BIM process will become fully integrated with these systems. You'll be able to optimize heating, ventilation, and lighting systems based on the real-time use of spaces.

Building a solid foundation

BIM can dramatically change many industries, and it needs a combination of people, processes, and platforms with data at the centre. Instead of the traditional industry resulting in one built asset, BIM will provide two, a physical and a digital one. Make sure that both are well designed and constructed securely using best practices, and that they're easy to understand and maintain during their use.

Chapter 3

Examining the Information Part of BIM

The projects you work on are packed full of data, from costs and quantities to certificates and standards. Every component of a built environment project, whether it's a building, bridge, tunnel, or airport, is accompanied by a wealth of associated performance measurements, values, and facts. Somehow you need to manage and maintain this information as the project evolves. Information management is a huge task and traditional methods can be very inefficient.

Chapter 1 includes our definition of Building Information Modeling (BIM), simplifying many of the alternative (complicated) explanations you may have heard into a clear, concise sentence you can easily remember. The most important thing to realize is that BIM isn't just a technology, and it isn't just about engineering geometry or fancy visualizations. You need to be able to understand the project beyond how the components fit together and how it will look when built. In this chapter, we take a tour of the information aspect of BIM and show why it's literally at the heart of the term BIM.

Comprehending What Information Means

People often describe BIM as a data revolution. "Why?" you may ask. People hype BIM to be many things and have wide-ranging impacts, but fundamentally it's a decision by a project team to change the way you share information, cooperate, and collaborate on projects. It creates value by demonstrating that you can work in more efficient ways. The most significant change is in how the project team manages information across the life of a project.

You can explain the idea of the "I" in *BIM* as information modeling to someone by using the example of a human body. If you wanted to, you could accurately replicate the geometry of the entire human body, so that you had a perfect 3D model of human anatomy (and these models do exist as study tools for medical professionals). However, so much information is missing from the model that represents a real person, including age, medical history, family history, occupation, lifestyle, and daily routine. In the same way, a perfect 3D representation of a project is still missing so much information, such as the execution of building works, when components were installed and their likely replacement time, warranties and certificates, estimated energy performance data, and so on. The geometry is only useful with embedded property data.

For a fascinating crossover of construction BIM and medical anatomy, check out Arup's Project OVE at `http://arupassociates.com/en/exploration/bim-trial-project-ove/`.

A lot of industries have applied information modeling and managing information and knowledge across projects. BIM isn't something architects or contractors invented. In the following sections, you can find out how information modeling is used successfully by other sectors and how you can benefit from the information aspect.

Noting other industries that use information modeling

In simple terms, information modeling allows clients, designers, builders, engineers, fabricators, product manufacturers, owner-operators, and users to understand an entire project before construction, refine the proposal to avoid errors, and generate efficiencies. They can then output a digital copy of the built project, interrogate it for key facts in the future, and update or expand it as necessary when things change.

Because of these efficiency benefits, the construction industry isn't the first to think that information modeling sounds like a smart idea. In fact, some companies have been using the concept of data analysis for decades. The following sections give some great examples of other industries using information modeling innovatively.

By modeling a project digitally in terms of its information as well as its geometry, everyone involved has the opportunity to access, influence, and interact with the same data for different reasons. One of the key aims of BIM is to group all the information about a project into just one virtual place, but doing so is a long-term goal.

To make the most of existing technology, you need to ensure that the information and systems you use are interoperable; in other words, you allow and encourage data exchange and sharing to take place across the team. To help you with this, you can work to international standards available. In Part III, we go into much more depth about the various documents and protocols around the world that direct information coordination.

Automotive manufacturing

The automotive manufacturing industry is responsible for revolutionizing factory production through the modern assembly line at Oldsmobile and the use of magnetic conveyor belts by Ford Motor Company. The automotive industry is now acclaimed as a key innovator of digital information modeling. The 3D model is used to refine the design through the product lifecycle. For example, Suzuki recently aimed to remove 1 gram from every component in its next car: a win-win for company and customer, because weight reduction would result in huge cumulative material cost savings and lead to increased fuel efficiency in the car.

The model no longer exists just in the design phases; it's used as a fabrication model too, and the use will only increase with the evolution of 3D printing and intelligent materials. Automotive manufacturing has benefited from imposed international standards for design and safety, and the utilization of standardized computer-aided design (CAD) platforms across the industry. One of the key lessons you can adopt from car companies is the importance of the client being committed to the adoption of new technologies that support the supply chain to ensure interoperable information.

The concept that expands an evolving information model into project management has been used for decades and is more accurately called product lifecycle management (PLM). Chapter 6 goes into more detail about *integrated project delivery (IPD),* a term you'll often hear in the same sentence as BIM, and IPD provides an analogy that's closer to PLM. BIM and IPD processes working together can be powerful.

Aeronautical and aerospace engineering

Aeronautical manufacturing has advanced to the point where every commercial plane is designed and built using a comprehensive information model. More than 20,000 global component suppliers and manufacturers can be involved in the supply chain for one aircraft, so the only way to manage and coordinate that amount of data is via a central hub utilizing live data. In the examples that we've encountered, the plane only ever exists in the virtual environment, and no prototypes or mock-ups are built for testing; the manufacturer does everything in a digital form until the final build, such is the trust in the data management.

Aerospace information models are also embedded into PLM systems, which completely integrate teams and information. This results in high levels of manufacturing quality and efficiency. Airplane manufacturers even embed checking approvals with qualification data, so that managers can track every decision back to an individual. If a component fails, airplane companies can see not only the use of that part in other aircraft and who manufactured the part, but also who installed it and what else they installed. You can see an overview of aeronautical PLM in Figure 3-1.

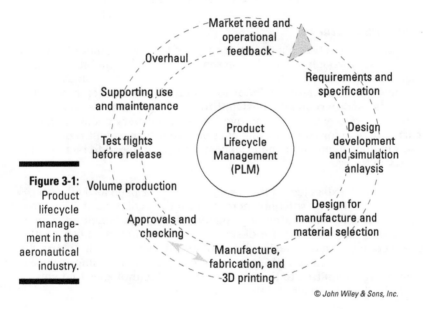

Figure 3-1:
Product lifecycle management in the aeronautical industry.

The industry needs to respond quickly to market demand and performance requirements such as fuel efficiency or new emissions legislation, and it does so on a global scale. One has to be able to zip large amounts of data around the world to allow the kind of international team-working required in aircraft production. As a result of this precise quality-assurance process, the concept-to-delivery time for the latest single-aisle passenger jet design is less than 18 months, and in 2013, Airbus delivered 41 identical aircraft every month.

NASCAR and Formula 1 racing

You may know that in the mid-1990s a NASCAR pit crew became synonymous with continuous improvement when Ray Evernham's Rainbow Warriors went through seasons of strength and agility training, video replay, and rehearsal to *choreograph* driver Jeff Gordon's tire changes like a ballet. Emblazoned with "Refuse to Lose" across their chests, they took lessons from professional football and gave specific roles to each person based on his individual skills.

Since then, the crews have gradually tried to shave seconds off the pit times through these same low-tech methods. Now, though, teams are using real-time location systems (RTLS) such as radio-frequency identification (RFID) tags to assess and train pit crews in practice, tracking the location and movements of the car, all the equipment, and the engineers themselves. The data instantly generates an information model to allow NASCAR crews to completely optimize performance and motion.

In Formula 1, where the cars don't refuel and only have a single wheel nut, in the 2013 season Red Bull Racing changed all four tires in 1.9 seconds. During the race, the Formula 1 constructor teams use real-time information models of the cars to understand every element of car performance, from tire pressure and engine temperature to aerodynamic effects in different weather conditions. Interestingly, in order to maximize analysis, teams often feed this information back to their factory headquarters and then back to the racetrack. Think about the speed of your home Internet connection and consider how fast the communications need to be to relay data transmissions back and forth during a race.

You can take another great lesson from NASCAR and Formula 1. In selecting pit crews, the racing teams evaluate the applicants, and through the right assessment can work out where their skills and weaknesses are. The teams then provide training tools particular to that situation, rather than putting everyone through the same training or making everyone learn to do every task.

Not everyone needs to know everything that's going on. Higher levels of management will want an overview picture, but don't need the details of your software training. At the same time, you can't expect someone with a very specific role in proofreading or object modeling to be able to explain your entire strategy. Think about what messages you need to relay to your whole team and what parts may only be relevant to specific people.

Also, some of the crew are former athletes, working in this new industry for the first time but using skills from their previous sports. In the same way, which of your colleagues can you see adapting to BIM and data management roles who can bring a different kind of expertise to the table? Where can someone add value to your organization's BIM implementation?

Eyeing the main benefit of information modeling

The main advantage of information modeling is quite simple: no one *owns* information. You may be familiar with the traditional view of the concept of *ownership:* "This is my steelwork design! Why should you have access to my drawings?" or, "If you really want, I can tell you the height of that door, but I don't see why you need it."

Every single person on a team has a responsibility to make his information open and accessible in shared locations. BIM is more than just an innovative tool that helps some of the design team and an integrated central data platform that the client requires you (and everyone else on the project) to use in an open and intelligent way. Eventually, at handover, the client or the eventual building owner can take ownership of all the native model and exported data. BIM is a lot better than receiving hundreds of boxes filled with paper.

Right of ownership of the model and the legal implications of BIM on fees, intellectual property, and copyrights are the subject of discussion in the construction law community, with a gradually increasing set of test cases for the courts. In Chapter 13, we explore these issues along with other challenges surrounding BIM, such as security, risk, and contractual liability.

For example, consider the construction industry, which has traditionally worked in *silos,* usually reflecting the job titles or qualifications of each team or office, such as architects, structural engineers, quantity surveyors, mechanical engineers, and landscape architects. Everyone has his or her own set of data, probably using different software programs and managing information coordination in-house, releasing just what the client requested or another team needed as the project progressed. This is in direct contrast to the idea of BIM, which allows everyone in the supply chain to add content to a unified model.

Information modeling is the future of construction

The construction industry is one of the last areas of global business to adopt standardization and refine information management processes in order to generate efficiency. You can argue that construction activities are manufacturing of a sort. Construction is the creation of an object made of many components; that object just happens to be a very complex project containing hundreds of thousands of advanced sub-objects like steel beams, timber, bricks, and concrete. Nothing prevents construction from leading innovation rather than playing catch-up with the manufacturing sector, but it needs to understand the power of information.

Using entirely digital information is a big step forward for built environment teams, but doing so isn't going to solve every existing problem in design, construction, and asset management processes. It can't ever be a silver bullet, because the success of a project still relies on people understanding that they need to communicate with other team members. One of the principles of BIM could be the following: information is only useful if you're sharing it. This principle should go beyond defined roles and job titles; if you see something is wrong, tell whoever needs to know.

Realizing That Information Is the Heart of BIM

Think of how you currently share information with others on a project. Perhaps your inbox is full of emails with attachments, or you receive paper documents via postal mail. Even when you access shared, digital information by using central drives or cloud-based storage, often the information is a certain kind or accessed only by a specific team.

Construction is full of data, but most people have never really made the most of it. For decades, in order to access data you needed to have a physical copy of it, on CD or paper, or digital versions, such as a PDF. Now multiple users can see the same data online at the same time. The industry needs to move from a mind-set of multiple users owning their data individually, to collectively maintaining one accurate and up-to-date version of the information, with many people accessing it whenever necessary. You can describe this as moving *from documents to data*. In a literal sense, information is the middle word in BIM, and you should think of information as the beating heart that keeps BIM moving. Building modeling has existed for a long time; the addition of consistent and open data to the mix gives BIM its power to change these traditional processes.

The majority of national BIM strategies in existence have made the focus on information deliverables clear. For example, the UK Government Construction Strategy BIM requirements combine the handover of information as a native BIM platform deliverable, plus a Construction Operations Building information exchange (COBie) database. It's vital that the handover isn't just geometric visuals, but also essentially a live database of project information.

Of all the information embedded in your virtual project, everybody needs access to at least one piece of information. Some need to run thousands of queries, and some need access to everything in order to coordinate the model. Information flows from concept to demolition.

Think about the project timeline and who needs access to the model at various stages. Ask yourself who are the generators, reviewers, and receivers of information. Here's a brief definition for each broad group:

✔ **Generators of information:** They are BIM users, such as the client or concept designers, who will be generating initial information, adding data as it becomes known, and continuing to improve the model by evolving existing parameters in the model. Information generation happens all the way through the project timeline.

- ✔ **Reviewers of information:** They are the users who need to make decisions to progress the design or construction work who will be analyzing the data already in the model and reviewing how to achieve the required levels of performance or to collaborate to avoid clashes in geometry. The majority of a project delivery team will be reviewers.

- ✔ **Receivers of information:** They are the end-users of the data in BIM, using either live project data or exporting the documents, reports, and drawings. For instance, the caretaker janitor of a public school or library may want to generate maintenance information from the model or pass information to the client.

You can begin to see how information is the heart of BIM when you see how the data evolves and grows as users add to it and how it will be interrogated and extracted by others across the project timeline. The following sections look at the generators, reviewers, and receivers of information in greater detail.

Adding information to the model

Generators of information are the briefing, design, and early development teams. These teams produce the fundamental information at the start of the project. Alongside the design geometry at this stage, think of the following.

Site information

The location of the site may be obvious, but it carries a huge amount of embedded information about the environmental conditions and quality of land. Geographic information systems (GIS) and BIM evolved independently, but they have a lot in common. You can find hundreds of examples where infrastructure contractors have successfully married the two to improve projects at a macro scale. You can use GIS to understand the effect of topography and your site conditions on a proposed development.

Outline and performance specification

Understanding the client's requirements for the project is a fundamental part of the design process, and without a good brief you can't begin modeling. Rather than thinking of the briefing and specification as two separate activities, use master specification tools to generate an outline specification as early in the project as possible and then use that record as a foundation for developing the full specification.

Say that a wall in your project has specific performance requirements (fire resistance or acoustic reduction) but the design team or client can't decide on construction type. Specification tools allow you to record the performance requirements as part of the building information model. You can then link a placeholder object (such as a generic blank wall) in the geometry with the relevant specification data. This is a great example of something you can't model with 3D CAD alone.

Planning code requirements

You can now find examples of where you can embed into the information model local code requirements like proximity to neighboring buildings or trees, limits on the height of buildings, and sustainability factors such as regional public transport routes. Doing so instantly increases your understanding of the impact of planning codes on your development, speeds up the design process, and can even help you justify your development to planning officials. We demonstrate some exemplar tools in Chapter 19.

Using information in the model

Reviewers of the model are detailed design and technical design teams and consultants, such as structural engineers and mechanical engineers working together with architects, lighting designers, landscape designers, and the wider supply chain. Consider that the whole project team could be involved by this stage. In the following sections, we help you see how reviewers of the model coordinate, collaborate, and use the data BIM provides to the project team.

Traditionally, people had their own set of information, and coordinating everything was difficult. In the long term, a future method called Level 3 BIM (or iBIM) will provide cloud-based environments for all members of the project team to access. In the meantime, people still have separate sets of data, but BIM ensures that they all speak the same digital language.

Assembling these interoperable models together is called *federation.* You may hear the term *federated BIM,* which basically means that the various parties have combined their work (fabric, structure, lighting, mechanical services) into one model, making it easier to see where the problems and clashes are. However, the authorship of the models is clear, so no confusion exists about liability or design responsibility. The information also has to go through a number of approval *gates,* which refers to a regular process of coordinating multiple sources of data into one, clear package of information for pre-defined stages of the project. Refer to Chapter 8 for more discussion about federated BIM.

Prescriptive specification

As the project evolves, you can refine in greater detail some of the objects that were placeholders at early stages. You can begin to rationalize the design and specification to indicate as many properties of each object that you know; for example, the appearance, materials, and finishes of the items that you'd previously specified just in terms of their performance. This is moving from a descriptive- or performance-based specification into a prescriptive one.

Manufacturer information

After the client makes decisions about systems in the project in the information model, you can begin replacing generic and placeholder objects with real building product manufacturer data. Alongside proprietary BIM objects that reflect the change, design teams can coordinate the specification data to provide access to manufacturers' properties like energy consumption and guidance such as operation or maintenance instructions.

Energy analysis

Another great benefit of information modeling is that you can clearly see the impact of design, orientation, and engineering decisions on the energy efficiency of the project. Consider that designers and consultants can run sophisticated simulations to demonstrate solar gain and shading, thermal mass calculations, and power consumption. Now combine that with what you know about the energy regulations in your area, including Leadership in Energy and Environmental Design (LEED) certification in the United States (www.usgbc.org/leed), Building Research Establishment Environmental Assessment Method (BREEAM) for assessment in the UK (www.breeam.org), SKA rating for fit-out projects in the UK (www.rics.org/uk/knowledge/ska-rating), or Green Star in Australia (www.gbca.org.au/green-star). Harness the power of BIM technology so that you can see easily how small changes can improve the assessment.

Whenever you're adding information into the model, think about what you need to embed in the model and what you can just link. In BIM platforms, you can associate objects with nongraphical information. Sometimes providing linked information is more useful.

Exporting information from the model

Receivers of the data are the *end users,* which means anyone who may want to export information into another format, print an image or PDF, interrogate the model in a BIM-viewer software program, or generate maintenance instructions, such as a client or contractor.

The quality of your output can only ever be as good as the information within the BIM. Especially in federated models, you're often reliant on the quality and accuracy of others' work. For BIM to really work, the entire project team has to trust the professional competence and integrity of all the other parties.

In the same way that you can add information to the model in various forms (geometric or text-based), you can also generate a range of outputs.

Here's a list of potential scenarios:

- ✔ You need to view part of the BIM for an internal design review.
- ✔ You need to output all data at contractor-tendering stage.
- ✔ You need to export the BIM to a model-checking tool to look for errors and clashes.
- ✔ You need to generate PDF drawings for building code approval.
- ✔ You need to print a visualization for a promotional marketing campaign.
- ✔ You need to send a drawing of a detail to a contractor on-site. Increasingly, contractors are beginning to use federated and cloud-based BIM for construction, and we look closely at on-site use in Chapter 16.

Information export often follows a pre-defined series of agreed outputs called *data drops*. The project's BIM protocol describes the required file formats and exchange schema, which depend on the purpose of the output. Don't worry; this just means a clear plan already exists for how the right amount of data needs to export and communicate your information, at the right time.

Maintaining Information in the Model

You need to think about the project beyond the traditional end of a design phase and past the construction phase and handover. In other words, knowing who updates and maintains the information is important.

You can think of BIM as a constant work in progress, because it refers to the whole lifecycle of a project. To realize the true vision of BIM, you want to be able to use the information not just during design or construction but also for facility management and operation, and potentially to understand actual building use. You certainly don't want to have to start the whole process again when the project later needs extension or renovation.

Consider the following questions and think about how the answers can help you to plan a maintenance strategy for your model information:

- ✔ When will your model go out of date? Will this happen as soon as something changes the design on-site?
- ✔ Will your model be accurate for as-built records? Will it contain manufacturer objects instead of generic ones throughout?

✔ How will you keep the model up to date? Would your client consider paying someone else, like an asset management company, to keep the currency of the model during operational phases?

✔ Will you need to link your model with other systems and synchronize the data you're using with other teams further into the project's lifecycle?

The following sections look at these issues of updating the model in detail and help you to recognize that BIM isn't a static output, but that it's an evolving flow of information.

Tracking the different information categories in BIM

Information in the model can take many forms, including the following:

✔ **Primary:** Like the geometry or a drawing, which the design team generates first-hand from BIM

✔ **Secondary:** Like energy analysis, environmental assessment, or a health and safety plan, which require review of the primary information as part of its creation

✔ **Tertiary:** Links primary data (like a ground-floor plan) with secondary information (like a fire strategy) to form outputs and documents specifically for building users (the fire escape map you may see next to the door of a room)

When primary information (the height of a wall, a detail of a window) has been altered, the model makes this clear, and different tools can compare models for precisely these kind of obvious, geometric changes. However, how do you manage the resulting update required to your secondary information? What if changing the height of the wall has changed your acoustic strategy, or the detail of the window has an impact on your energy analysis? How would you stay informed?

Many of the BIM protocol documents, in particular the CIC BIM protocol, require the appointment of an information manager. The manager's job is to keep all the information flying around synchronized and coherent, and usually it falls to the project lead or a similarly senior team member to take on this chief coordination role.

The protocols also suggest a common data environment (CDE) to help the process of creating, sharing, and issuing production information. The CDE is simply a place you can put all the information about a project and share it with everyone. The CDE isn't just for geometric models either — oh no. Guess what else should be in there? Information like documentation, registers, and schedules.

In Figure 3-2, you can see how the CDE brings together information from multiple disciplines on the project team. The information from each sector could arrive as separate models, at every project stage. Combining and coordinating the shared information is vital to the success of BIM.

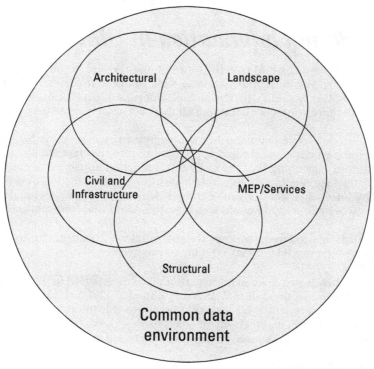

Figure 3-2:
A common data environment.

© John Wiley & Sons, Inc.

You can implement the CDE process in different ways:

- ✔ A project server
- ✔ An extranet
- ✔ A file-based retrieval system

A CDE has some great advantages, including the following:

- ✔ It provides a single source of information (more like a single source of truth!).
- ✔ Shared information reduces time and cost by providing coordinated data.

- ✓ Members of the project team can generate multiple documents or views from different combinations of the central model files.
- ✓ Spatial coordination becomes a simple by-product of the CDE.
- ✓ It delivers production information right the first time.

Updating information in BIM

You may want the BIM to represent not just the end of the design phase but possibly the as-built information as well. Doing so involves replacing the construction objects with the as-built information, including all the chosen manufacturer objects containing all their relevant properties.

Traditionally, each member of the design team was responsible for keeping his own data current and applying the relevant changes and updates as required (assuming that the individuals reached appropriate contractual and financial agreement). If your role is designer, you're responsible for maintaining your design work; if you're the services engineer and a change to the ductwork occurs, you could be asked to update that information.

The information manager may well be required to coordinate the development of the model long into the life of the project.

Often, this desire to keep information up to date is tied in with liability and design responsibility. With every object in BIM, liability is implied in a number of properties. It's vital that you keep this information updated so that it's accurate when somebody needs it further down the line. For example, consider these questions if a product was to fail in use on a project:

- ✓ Who designed it? (Who is responsible for changes to its design?)
- ✓ Who manufactured it? (What if that manufacturer goes out of business or discontinues that product?)
- ✓ Who installed it? (Could another installation company carry out the same job, or does the manufacturer have specialist knowledge or equipment?)
- ✓ Why did it fail? (Is it a design fault in the product, a mistake in how it was used in the project, or poor installation?)
- ✓ How was it fixed? (How is the audit trail recorded? If something goes wrong after the fix, who is responsible for further repair or replacement and the associated cost?)
- ✓ When was it fixed? (How often has this happened? Are these products failing regularly?)

✔ How long will the warranty last? (Is the product still in warranty? Can the manufacturer or installer guarantee the work for a minimum period?)

✔ When will it need to be replaced? (Is it more cost effective to regularly replace an entire set of a repeated item and test accordingly or to repair and maintain on an item-by-item basis?)

Synchronizing with other data using BIM

The Avanti Project, which we cover in more detail in Chapter 6, was an initiative by the UK industry to demonstrate performance improvements through collaborative working. In 2007, it found that intelligent information modeling accompanied by relevant standards and procedures saved 60 to 80 percent of the effort spent hunting around for information and documents.

Irrelevant of where in the process you're personally involved, in a BIM environment a number of different members of the project team could use your information at any stage of the project, including operational use and retrieving information at a built asset's eventual demolition. That's an important fact to consider when you're rushing to meet deadlines and finish the design work.

Going a step further, when the Royal Institute of British Architects (RIBA) developed its recent update to the *Plan of Work* (www.ribaplanofwork. com) — a guide detailing tasks and outputs for the project team across eight work stages from Strategic Definition to In Use — it adopted a circular logo, indicating that the process isn't linear. Instead, think of all the information from in-use projects feeding back into the next strategic briefing stage.

So in fact, the corporate knowledge of the design team should always be making your next project even better.

One of the terms you may hear when discussing BIM exports is XML. *Extensible Markup Language* is simply a way of ensuring that both you (a human) and the computer you use to run a BIM software platform (a machine) can read the formatting of textual information that your model produces. More importantly, it means that you can have multiple software programs that use the information from BIM and talk to each other by using a common schema. In Chapter 9, we take a look at XML, Industry Foundation Classes (IFC), COBie, and other means of data exchange in greater detail.

Achieving efficiency

The key goal of BIM implementation is to demonstrate that improved access to project information and a collaborative environment saves time, removes iterative effort, and has vast economic benefits for every single person involved.

The banking sector took standardization and applied it across an entire global industry — developing the standard size and magnetics for a credit card and releasing hardware and software requirements for how to read the information in a consistent way. Can you imagine every bank having its own shape and size of credit card? "Oh, this ATM only takes square Bank of Dummy cards, not rectangular Bank of the Universe cards. I think there's one down the street."

Governments have often looked to other industries to encourage greater collaboration in construction. For example, in the UK the reports "Constructing the Team" by Sir Michael Latham and "Rethinking Construction" by Sir John Egan recommended change in a fragmented industry through concepts like lean construction and partnering and integrated project delivery. They've had varying levels of success, notably in developing new forms of contract.

The difference with manufacturing industries is that they were pushed to find new efficiencies to maintain a competitive advantage in global markets. In order to do so, they imposed international standardization and set clear processes across every part of multi-faceted companies. You can learn from this; they didn't just recommend new methods or advocate change without an incentive or driving factor.

The UK government could have just suggested BIM targets for 2016 as a means of delivering increased value through cost, energy, and delivery-time savings, but instead it chose to define clear objectives for project procurement and full collaboration that companies who want to work on public sector projects have to meet from that date. That's the incentive, and it'll encourage the private sector along too.

Chapter 4

Discussing the Modeling Part of BIM (and Management Too)

*T*his chapter takes a look at the last letter of the acronym BIM, the M. In most definitions of BIM, the M stands for *modeling,* and we agree with that, but only because modeling can mean a lot more than just geometric modeling, using computer-aided design (CAD) software. A building *information model* can let you navigate and experience your project in an immersive, three-dimensional (3D) environment. The majority of platforms also let you build digitally in 3D space and produce the 2D drawings and photographic renders you require. In the model, you can represent every component and object in the real project. The more precise you make the detail of your modeling, the more construction and coordination accuracy you can extract when you need it.

However, a lot of confusion around BIM comes from a simple misunderstanding: that BIM is the same thing as 3D CAD and that the modeling part involves just the visual aspect of seeing the objects on screen. In this chapter, we help you to understand how BIM makes modeling built assets and construction projects easy.

Some people say that the M in BIM stands for *management* instead, and in this chapter we also look at how the model develops across the project timeline and how the project team maintains and updates the content through good management.

Grasping What Modeling Means

What do you think of when you see the word *model* in a construction sense? For a really long time, designers, architects, and construction teams have used hand-built models to describe what a built project will look like, from houses to cathedrals to citywide developments and bridges. These kinds of models are a great way of visually representing how a space will be affected by what's proposed. As students, we spent hours and hours building these in card, paper, balsa wood, and glue. Now, 3D printing and laser-cutting methods allow people to assemble accurate models very quickly.

But that's only one kind of model. The word *model* can describe anything meant to be a representation of the real thing. You may have heard of environmental or weather modeling when scientists predict the impacts of climate change, or financial modeling to understand the patterns found in stock exchange and currency markets. Biological models are used to represent the effect of change on humans or nature. From a construction perspective, information models are about representing all the data in a built project, including how it looks.

If you're a fan of word meanings, here's a great fact for you. The word *model* comes from *modulus,* which means *standard* or *measure* in Latin. People have used the word to describe a scaled representation of a building since the 1500s. It's the same derivation as modular building, where standard blocks and parts, even small building types, are constructed together.

So model in the sense that this book uses it, in the term Building Information Modeling, isn't just about a tangible model nor is it just about the physical size and shape of a project. It's an analogy, a representation of the real thing. If you want to describe a building or a bridge, you can model it, but you can also model a team's performance or the results of an election. Models can represent every aspect and impact of a project or event on its location and its users, especially the information.

The important thing to remember is that the beauty of a high-resolution visualization is only skin deep. Without accurate construction details and embedded data properties, the model is only for show. The difference between visualizations and a high-quality information model is how much useful data can be extracted from the model over the course of the project.

Often, good information models also contain a high level of accurate geometric information, forming an advanced 3D graphical representation of the project, just like the visualization in Figure 4-1.

Figure 4-1:
Manchester
Central
Library by
Ryder
Architecture.

Illustration courtesy of Ryder Architecture

In the following sections, we demonstrate some of the benefits of digital 3D modeling and also clear up a common misconception about BIM and 3D CAD.

Going digital in a changing industry

The construction industry is changing through the adoption of digital technologies, and modeling is no different. A practical shift has occurred from manual methods to digital ones, like taking a 3D object on screen and translating it into printed forms. But a move has also occurred away from paper and book-based data to electronic information using the power of the Internet and, most recently, the cloud to compute and model different scenarios and options for built environment projects. We talk about BIM in the cloud in more detail in Chapter 7. Also, Chapter 19 looks at the future of BIM, including how technology might change information management.

Recognizing the benefits of modeling

Even before we start thinking about what BIM can do, there are huge benefits just in the native power of 3D design platforms like Revit, McNeel Rhinoceros, and Gehry Digital Project. CAD modeling has evolved for many years to create amazing things.

Computer-aided design (CAD) and computer-aided design and drafting (CADD) are terms that have been around in the industry for decades. The terms are pretty much identical, but one deals with outputs as well as design. Although there have always been add-on applications and databases that connected with CAD software to link the graphical information with more intelligence, this isn't the same thing as BIM.

Here's a list of three key benefits of 3D modeling:

Making design easier

These tools have moved building design forward, allowing you to construct forms and building types that would have been impossible without them. The technologies allow for innovative construction development and incredible visualizations. You can also begin to use sets of rules to automate some checks on the model; for example, clearance zones for maintenance or minimum distances to meet building codes or regulations.

Making coordination simpler

The parametric functionality of modern software means that the changes you make are displayed in three dimensions and in all views of the project. You can switch views and understand the impact of your design decisions really quickly. So much of the industry is stuck in two dimensions, when in fact 100 percent of what you design ends up in three dimensions. BIM is just as much about 2D as 3D, because your outputs may be 2D, but working in three dimensions makes coordination so much more accurate.

Not just that, but many BIM platforms now have the ability to coordinate multiple files and run things like clash detection to assess differences between content, allowing many users to collaborate on the same projects. By using template files, the project team can exchange standard outputs for scales and line weights/thicknesses between multiple offices so that outputs are consistent too.

Analyzing energy use

Now designers can use software platforms like Ecotect or IES to ensure that the energy strategy for the project is right. From running solar studies for passive heat and ventilation or understanding areas of shading at a building scale, 3D tools also can model electrical and mechanical engineering setups for citywide energy developments. In combination with intelligent energy software, engineers can use these tools to understand routes, peak loads, issues, and risks.

Because one of the foundational aims of BIM is to save money in the operational phase of projects, designing out wasted energy and refining the energy performance *now* can be cost effective in the long term.

Proving that BIM isn't just CAD or 3D CAD

The biggest misunderstanding in BIM is that if you buy one of the major modeling tools like Autodesk Revit or Navisworks, Graphisoft Archicad, Bentley Microstation, or Tekla BIMSight, then that's it — you're doing BIM. The vendors of these platforms often advertise that you can produce BIM projects just by purchasing their product. That's not just misleading; it's the opposite to the real objective of BIM, where multiple streams of data converge into one river. Think about what outputs are generated by different parts of the industry, different disciplines, and at different stages in the project, which can help you realize how one CAD software can never support coordinated BIM all on its own.

A lot of people think Revit is BIM. We're not in the business of trying to undermine any particular platform, but you have to realize

that Revit is just one of many BIM tools. Put simply, you can't say the word "BIM" when you really mean "Revit." The reason is that you could produce a 3D building in Revit without embedding anything other than geometric information in it. In reality, you couldn't build the physical asset without referencing other information, from environmental performance calculations to building product manufacturer vendors' data.

So here's a confusing sentence: What Revit can produce isn't BIM; it's just part of BIM, but Revit can produce *a* building information model. Does that make sense? A lot of tools can produce building information models, but there's no such thing as the Ultimate BIM Tool that fully integrates everyone else's information into one perfect utopian model. At least not yet!

Confirming BIM Is a Process

If you're cooking, you take a range of ingredients from your kitchen and, through preparing, chopping, mixing, and blending, you produce a meal using various equipment. You may be following a recipe to use other people's experience. In big kitchens, many chefs work together.

Think of BIM in exactly the same way. No one magic ingredient or one piece of software works for an entire BIM project. BIM is a process and it's also a skill to learn, just like cooking. BIM also has recipes, like standard documents and protocols, to help you get the best results.

You'll no doubt meet a few experienced people who say that BIM has been around for a long time, and vendors and sales companies (and *For Dummies* co-authors) are using BIM as a buzzword to cash in on something that's existed since the dawn of computing, as demonstrated by the manufacturing industry.

Unsurprisingly, we strongly disagree! Not only have the tools to make BIM work in construction only recently become possible, but the various industry representatives, organizations, and heads of big business have never been brought together in a coordinated way to agree on the protocols, processes, and conventions needed to assemble BIM-ready teams and understand the frameworks involved. They needed to sort out a lot of disagreements first.

Here we show you that the model isn't just a one-time output as a snapshot of a moment in time; it's a living, growing process model across the project timeline that can be interrogated, adapted, split, and combined many times over. The following sections look at the process of modeling and how you can move from traditional, drawn information into 3D processes, and then update the model information to keep fluid information accurate.

Modeling your drawn information

If you're used to working in 2D environments, the jump to 3D can seem like quite a chasm to overcome, and not just graphically. Seeing the development of the model as an ongoing workflow and not independent drawings as individual pieces of work is important. The drawings required at a particular stage are one of many outputs you can take from the information model.

The amazing functionality or powerful capabilities that a particular 3D software has don't matter; people use software to produce what they're familiar with and what they think the users of the outputs actually want to see. So you more than likely encounter people using 3D software but not really working in three dimensions, or people fudging the accurate construction information to get the visual output they want.

If you're making the transition from standard CAD software too, realize that you follow a process: first discovering how to model in 3D and then understanding how BIM processes impact your work. These two tasks are separate, and you can't confuse them.

In a 2D environment, your main concern is the graphical output and having a consistent approach to the line weights and thicknesses to convey meanings, such as wall edges or hidden pipework. These standards for 2D outputs are still important in BIM projects, but are less vital because multiple other ways exist to interrogate the model information to find that kind of data. It's more important that the 3D content is enough for consultants, contractors, and users to extract the real information they need.

Always model accurately. Don't take shortcuts. Not only do you have the best chance of avoiding clashes, but you also make sure you resolve every connection, joint, and opening in the digital realm. In addition, you locate buildability and safety issues ahead of time. So zoom in to check how things join.

Changing the model and changing things on-site

During a BIM project, the amount of data in the model increases from the concept and design phases to the construction phase, so a contractor can use the model for on-site instruction and specification. Then, as the project progresses, the model evolves to match final decisions made by the client and determined on-site. This evolution is sometimes called *level of development* or *level of definition,* and it's split into two parts:

- Level of detail (LOD) (geometry/objects)
- Level of information (LOI) (specification/data)

The critical thing to understand is that the client may require different levels of detail and information for the same system at any given time. Refer to Chapter 10 for more information.

The information model isn't the same as a fabrication or manufacturing model. We've seen examples of models that practically ground to a halt in standard software platforms running on office hardware, because they contained too much detail, as if every object was going to be milled directly from the model. At every stage, ensure the model contains all the information that you need, but nothing that you don't require.

One of the most difficult concepts to understand is how to keep the model up to date, but doing so is super-important. The model goes through a number of stages in order to represent the most accurate set of data at the end of the project. In particular, when things change on-site, it's really important that the contract has determined who is responsible for updating the information in the model, so that the model becomes an *as-built* representation of the project, not just the *design intent.*

Updating the model shouldn't happen just at the end of construction either. Because some construction phases are very long, you want the model to always be the most accurate and current set of information possible. Therefore, someone should be in charge of ensuring this is a constant process throughout the project. That's where you need model management.

Managing the Model: Federate to Collaborate

Model management has two main objectives: to ensure currency of the data in the model and to coordinate information coming from across the project team and fitting it all together. You won't be able to find a one-size-fits-all solution for the perfect BIM tool. At the moment, it just doesn't exist, because what a client, surveyor, architect, engineer, landscape architect, contractor, interior designer, operator, and caretaker need to input and extract is too different to be handled by one model.

You can find out how to combine information from multiple sources in the following sections. Later in the section, we show you how this incorporating of models using open source methods encourages collaboration.

Federating the model

There's a way around the lack of a one-size-fits-all solution to generate one perfect model that suits everyone's needs. It's called *federation,* which basically means joining together small pieces to make something bigger, just like a sports *federation* is made up of individual clubs. You can think of the federated model as individual models, usually divided up by discipline (the architect's model, the mechanical engineer's model, the structural engineer's model, and so on), but linked together so that they can communicate. Federation is the foundation of Level 2 BIM.

Evaluating the benefits of federation

Federation has two important benefits:

- Federating multiple models provides the opportunity to appoint a gatekeeping role, to manage and plan the process of combining each individual source into the federated model. This level of control is useful in comparison to trying to manage information being added by multiple users in one virtual space.

- From a data security and copyright point of view, you can clearly see where each part of the model has come from. Everyone can retain ownership of her intellectual property, rather than the collective team drawing everything in one space.

You can easily get into a lazy way of thinking about construction information that's existed for a long time: that something lives in just one place. For ages, the model has been the place to put graphical information and the specification, nongraphical information. This is still the case, but now you can read, reference, and report that data from anywhere in the federated model. You can choose to produce outputs from the live model or extract just the information you need and use plugins and add-ins to ensure currency of data. Just remember, a change in one place is seen as a change everywhere by all users of the model.

The various BIM protocols go into federation in more detail. For example, look at the BIM Addendum (`https://consensusdocs.org/Catalog/collaborative`), which states that models don't lose their identity or integrity by being linked, so that a change in one part of a federated model should never change the other models. This means that everyone's information can remain accurate and updated, but that an individual user can clearly see issues and clashes.

Highlighting a tricky part of managing federation

One thing that can cause a bit of a headache during federation of model information is that software vendors often release new versions of their platforms, sometimes major releases on an annual basis. Throughout the development of any software, vendors can also release new functionality or bug fixes as updates for you to download or to patch your current installation.

It's absolutely vital that everyone working on a project is using the same version (and build number; for example, Version 2.5.6_05112015) of a software platform. First, doing so ensures that everyone's installation behaves in the same way, avoiding crashes and giving everyone access to the latest tools. Much more impacting is the fact that most software won't allow you to open a file created and saved in a newer version of the program. So if any one user is ahead of everyone else, that user's information may become unusable for the rest of the team.

Ensuring that everyone is using the most up-to-date version of a software platform may make sense, but good reasons exist for not doing this. Think of the financial implications and practical reality of installing brand-new versions of software across a whole project team. What's important is that you agree at the start of a project with everyone involved what the protocol for software updates is going to be. Ensure that everyone is using the right version and agree what happens when a vendor releases an update.

Think of the model as being a central location where all the project information (whoever has produced it and wherever it has come from) is brought together and updated, and think about who is managing that process.

It's really important that you agree on standardized outputs with the rest of the project team and ensure that everybody understands the way the project team is going to use, exchange, and federate information, or you'll run into problems quickly.

Collaborating with Industry Foundation Classes and a common data environment

The goal of federation is to improve collaboration between members of the project team. For a number of reasons, the tools and software you use to model your project with shouldn't matter. In particular, *interoperability* — the ability for multiple software products to talk to each other — should mean that other users and project team members can use your information. Sadly, this still isn't the case yet. Interoperability and exchange formats are improving all the time, but they're not quite there.

The sooner you can become familiar with two terms, the better. These two acronyms describe ways to exchange data from different software and multiple sources as part of a federated approach. They're as follows:

✔ **Industry Foundation Classes (IFCs):** IFCs are rules for exported information, similar to the way MP3 and PDF are standard ways of exporting data produced in loads of different ways. The IFC has had a lot of teething problems; like a lot of good ideas, it's taken a while to find its feet. But the latest version is the industry's best attempt at everyone speaking one language.

✔ **Common data environment (CDE):** Think of the next step beyond federation as bringing together everyone's information into one virtual space. For now, CDE are usually just file-sharing sites and document management systems, but the utopia of the CDE is to have a coordinated, digital location, not unlike the World Wide Web, the pipework that powers the Internet. The WWW standards were a set of principles required before browsers could read any page. In the same way, it's the job of those designing the frameworks for BIM Level 3, which looks at coordinating BIM information in a true CDE, to work out the detail of how to build that utopia so that everyone has the same information at their fingertips. We explain the detail of a CDE in more detail in Chapter 8.

Chapter 5

Grasping the Fundamentals and Understanding What BIM Is

In This Chapter

▷ Recognizing what BIM is

▷ Cooperating and sharing information

▷ Changing the way projects exchange information

▷ Exploring the benefits

*I*n this chapter, you gain a simple overview of what BIM is trying to do, what it's for, and who benefits from it. We present the first things that you need to understand about BIM, what the point of BIM is, and what you need to know about the supply chain. We explain all of these issues, dealing with efficiency, financial, legal, and information requirements.

Figuring Out the Why of BIM

Everything you do should have a clear purpose with defined outcomes, and BIM is no different. This aspect is important in your BIM journey. People unite around a meaningful purpose, and you're more likely to succeed if you answer the why of BIM. For instance, you shouldn't collaborate for the sake of "you know it's a good thing to do," because the response would be ineffective and lead to either overcollaboration, which means waste, or collaborating on the wrong themes. So, BIM needs to have a defined intent. BIM effectiveness requires project teams to have a sense of common purpose if they're to commit to the direction of digital transformation.

You may recall the story of a NASA worker who was sweeping the floor when someone asked what his job was. He answered that it was to put a man on the moon. This was a clear purpose that John F. Kennedy communicated in 1969. This story illustrates the power of purposefulness as a key enabling tool. The worker had clear purpose in what he was doing. He wasn't doing some mere cleaning task; he was helping to put a man on the moon.

The following sections examine the point of BIM, which may be different for different people. Understanding how it can benefit you and your organisation is a vital key starting point on your BIM journey.

Examining the point of BIM

Having a clear and mindful vision is the starting point in your mission. The point of BIM is personal to you and depends on if you're a client, designer, or main contractor. Start with the end in mind and consider what the desired states you want to achieve look like. Determine where you are now and plan the journey. How you are going to get there? Then go do it!

So what's the point of BIM? There is no one standard answer here. Each path is unique, and the benefits realization can only be determined by working with your team and your customers to determine what good looks like and where you can make meaningful improvements. Don't create models and data just because you can. Ask yourself: What can BIM working and project data do for you that conventional working can't?

BIM helps you achieve a wide range of benefits for what you need, whether it be exchanging project data more collaboratively with little waste or actually giving your customers the assets they want that are efficient to operate and maintain. Refer to the later section "Presenting All the Incentives of BIM" for more discussion about the specific benefits.

BIM provides the information foundation for integrated teamwork, driving added value and removing waste from the design and construction process. The information provided by BIM is a readily accessible source for the teams involved in operating, maintaining, and adapting facilities. How you unlock the value of data and these tools is where you and your imagination kick in!

To BIM or not to BIM?

Perhaps you have the same feeling about starting your BIM journey that Hamlet had when he bemoaned the pains and unfairness of life but acknowledged that the alternative could be worse. We hope, though, that the thought of going digital is somewhat less dramatic. The decision to BIM should be clear cut and made for the right reasons. You may do BIM because your client wants a BIM deliverable, or your competitors are doing it, but fundamentally you make the decision because it's the right thing to do.

Don't do BIM if you can't work out the value proposition you want to set free. Creating data and sharing it should have real value. Building a robust BIM strategy and complementary investor case is a must. Plan for your journey and go for it when you and your leadership are committed to making it happen. It's like going in a diet: you know when you're ready to go Lean.

Having good data to measure success against is important, as is supporting the project teams in their delivery. However, if you haven't unlocked your goals, it becomes a harder decision. Check out www.dummies.com/cheatsheet/bim for a goal-setting worksheet to help you figure out your organization's goals.

BIM requires a long attention span, so be realistic with your goals, especially on your first two projects. Digital transformation usually takes you through three distinct stages:

- ✔ **Digital competence:** Figuring out new technology-related skills
- ✔ **Digital literacy:** The ability to understand and use digital information in multiple formats with meaning
- ✔ **Digital transformation:** Renovating business vision, models, and investments for a new digital economy

After you reach the latter stage, your BIM usage should enable new types of innovation rather than simply enhance traditional methods. After you reach this stage, you've long forgotten the "To BIM or not to BIM?" question.

Despite different definitions of the concept of BIM, you should consider the importance of the *why* of BIM over the *what*. In following sections you examine how you can add value and define data.

Categorizing BIM

Rather than trying to define what BIM is, some countries, such as the UK, have explicitly categorized different levels of BIM maturity, which are achievable through a series of different process, tools, and standards to achieve various outcomes and benefits realization.

Although the BIM concept may have many definitions, BIM should always boil down to a process of value-creating collaboration through the lifecycle of an asset, underpinned by the creation, collation, and exchange of shared 3D models with embedded or attached data.

A lot of misconceptions about BIM exist, especially around the technologies. Even though the software and increasingly the hardware, such as immersive glasses, are all pretty cool, they need to support new Lean ways of working to make any transformation. Digital retooling should support the business needs; the processes shouldn't be driven by your choice of software.

Focusing more on the *why* of BIM as opposed to the *what* is important. Be mindful that BIM isn't just a 2D to 3D shift, and that transformation only occurs when your digital usages enable new types of innovation and creativity, rather than simply enhancing and supporting the traditional methods. Essentially, BIM should help you realize planned disruption, whereby you can create a better way to meet a fundamental need with added value. Turning data into meaningful insight is fundamental to this process.

Adding value

BIM is as much about refining behaviors and processes as it is about retooling with some new software. Think about where BIM can bring you the most added value, map out the process, show the waste, and determine how your existing processes can be done better and faster using an interplay of digital tools and integrated processes.

You can and should apply the changes to your processes at all stages of the project lifecycle, not just when you start designing. BIM is at its most effective when you apply it right at the outset of a project. At that point you use data to help inform strategic decisions for a client, such as site selection, rapid energy modeling of existing assets to work out whether they need upgrades, and generally cycles where different design alternatives or schedule scenarios can be quickly simulated. This method is called *rapid optioneering,* and you can make more meaningful decisions earlier in your process.

Defining data

Good-quality and validated data is central to BIM, irrespective of how you define it — lifecycle data is the golden thread. Data, and especially the analytics, allow everyone to participate in the BIM process, from your CEO to the skilled technicians creating the data. Your construction data has real value and helps you improve delivery performance, reduce risk, and win work. However, in most cases the nongraphical data is your trusted friend in helping you answer queries such as: Does the design meet the accommodation schedule? What is the amount of embedded carbon within the design? Can you afford to build it?

As the project progresses, the data footprint progressively grows as you move from inception to handover, and indeed continued growth in the operational stages. The rate of this data growth in the models must be driven by the amount of information needed to answer each key question. Any more and you incur waste; any less and you can't effectively answer the questions. These questions are often known as the plain language questions (PLQs). PLQs are questions that the employer asks the supply chain to inform decision-making at key stages of an asset lifecycle or project.

Putting too much information into the model too early may

- ✔ Hamper both the design and supply chain options for others in the team.

- ✔ Lead to other forms of waste with the integral costs incurred by both the supplier and the client.

- ✔ Place constraints on the supply chain to offer compliant alternative designs, offering better asset or value performance.

Put simply, it's a bit like the Goldilocks enigma of too hot (too much data), too cold (not enough data), or just right (enough to answer the appropriate questions at that stage).

Try to avoid *infobesity*. This isn't a medical condition, but the act of putting too much information into the geometric model. Instead, think about relational databases and link your models across to other data sources. For instance, you may want to map across to a cost database rather than put the cost into the model, to avoid commercial sensitivity and for ease of updating.

Here we're talking a lot about data, either in the form of 3D geometrical data, nongraphical data, or indeed federated data. After you can fully understand this concept of defined information needs, validated outputs, and digital data transactions using information exchanges, you're a long way to understanding BIM.

Essentially, the data, the digital tools that support BIM, are the easy part. BIM shouldn't be solely a technical conversation; it should be about behaviors and how you can use the data and tools to bring people together. BIM should help people collaborate around a virtual campfire, where the various multidisciplinary parties can use the models to help understanding of often very complex assets and their systems. If you can bring people together and start a good conversation, you're onto a winner, and soon you'll be creating more innovative and integrated processes that yield better outcomes.

BIM isn't a single-player sport — the more participants sharing their information, the better the effect. BIM with just one organization playing isn't much fun and has limited value.

Thinking about virtual design and construction (VDC)

Yep, just when you got your head around BIM, we go and introduce yet another acronym: virtual design and construction (VDC). Effective utilization of BIM allows you to virtually produce a digital model of an asset, such as a building, in both the context of 3D object-based models and lots of associated nongraphical data. VDC then examines how you can use BIM to optimize the project by encouraging early decisions and engaging all members of the project team in an integrated value-engineering process.

Integration and innovation go well together, and VDC helps make that come about. The VDC approach lets you exploit BIM to simulate the delivery process, ensuring that you can accomplish an integrated approach to your project lifecycle. Get the project team in the same room, get your models on the screen, scrutinize them together, and watch the magic happen.

VDC was pioneered by the research at the Center for Integrated Facility Engineering (CIFE) at Stanford University. VDC is the management of integrated multidisciplinary performance models that includes the product — for example, facilities, work processes, and organization of the design-construction-operation team — in order to support business objectives.

The researchers note that VDC allows a practitioner to build symbolic models of the product, organization, and process (POP) early, before a customer makes a large commitment of time or money to a project.

Essentially, VDC looks at the virtual production, analysis, evaluation of the design, construction, and operation of a building and associated infrastructure in a digital environment. By working in a digital sandbox, you can replicate and understand how you can optimize the delivery solution. In essence, you're looking to build twice!

First, you virtualize the build process in the safety of a computer environment, running various scenarios as to how you best realize the desired outcome. For instance, what happens if you employ two tower cranes versus one? You can simulate and play out how the scenarios are likely to be realized in reality. Then you move from a virtual building process to a physical construction process that you can deliver with confidence, knowing that it's been tested and validated. Building better before built.

VDC brings together BIM and Lean

Virtual design and construction (VDC) encourages the amalgam of the two disciplines of BIM and Lean, using Lean construction and the Lean project delivery system. Through realistic modeling processes and simulation 4D, you can accurately examine and evaluate different possible construction processes and sequencing.

Lean construction is a version of manufacturing production techniques when applied to the construction industry using techniques that are aimed at both maximizing value and minimizing waste in the project lifecycle. Meanwhile, a *Lean project delivery system* is a model for managing projects in which there is a clear project definition represented as a process of aligning "ends, means, and constraints."

A big part of this Lean agenda is how project teams work together and communicate, encouraging co-location of the disciplines working on the project and integrated, concurrent engineering meetings where participants can effectively review, analyse, and refine the models and data in a simultaneous manner, rather than the traditional linear and time-consuming approach.

Defining what BIM isn't

Many descriptions exist of what BIM is, and in many ways the definition depends on your point of view or what you seek to gain from this digital approach. In most cases, saying what BIM *isn't* is easier. BIM is not

- Something that you can buy off the shelf or in a big box
- Just a design tool
- Just pretty 3D geometrical models
- Just the latest and greatest technology applications
- Just for buildings
- Just for architects
- Only for technicians, BIM consultants, and those younger than the age of 21
- Something that will only be available in a few years (go do it now)
- Hard (really!)
- Lonely (well, it shouldn't be; exchanging information with others is crucial)
- A cure for the common cold (don't believe all the hype or *BIMwash,* an inflated claim of using or delivering BIM)

Shaping Cooperation and Shared Information

The construction industry isn't known for playing nicely. Everyone lives in their silos, and in most cases everyone makes their profit from change and not by sharing toys. The industry needs to change if everyone is to achieve project success and improve.

BIM has made the industry hold a mirror up to itself. The industry has witnessed a landscape where construction disciplines work very much in isolation, concerned only with their part of the project puzzle and focused on outcomes. BIM offers a future with more connections and more data to help create a truly collaborative project team.

Collaborating and cooperating

Cooperation is a key tenet of any successful project and is a must for successful BIM execution. The intelligent clients realize this and create a framework to foster collaborative working that aligns with both their project needs and strategic objectives.

A project necessity is having a proper strategy to ensure meaningful cooperation between all the players. You shouldn't be collaborating simply for the sake of it. A proper strategy is all about the purpose. Collaboration is no different; it should have clear goals as to what cooperative working wants to achieve.

To begin to cooperate and collaborate, keep these points in mind:

- ✔ Consider with whom you want to collaborate — customers, suppliers, partners. Each has different implications for your approach.

- ✔ Identify what tools and frameworks you can employ in the project management system to support collaborative working and sharing of project information.

- ✔ Examine your practices. Doing so probably means introducing some new required processes such as relationship management measurement.

Here are some other areas you need to contemplate:

- ✔ A collaborative working policy statement for the project

- ✔ An overarching description of how the team members will work collaboratively

- ✔ Definition of what teamwork will do at corporate and project level to support the policy

- ✔ Provision of a framework and structure to publicize the vision, the expected outcomes, and the ability of the team members to collaborate

- ✔ Coherent thinking encouraged through regular multidisciplinary meetings

- ✔ Increased awareness of the importance of collaboration

- ✔ Information you want to share (saying you'll share everything is easy, but hard to follow through on; the right amount of appropriate information at the right time should be your goal)

Essentially, BIM implementation is 75 percent behavioural and 25 percent process and technology. Wanting to share information is a behavior that you need to foster on any fruitful BIM project.

The foundation is having a common data environment (CDE) to facilitate the collection, management, and dissemination of model data and project information between multidisciplinary teams in a managed process. The CDE provides a means of achieving a collaborative working environment where all parties can access the same information. The CDE is a requisite in achieving a cooperative project framework, and this also extends to non-BIM projects.

If the CDE is to be a success, all project team members must adhere to the agreed processes and procedures, especially naming conventions.

Generally, you can think of the CDE as having four distinct phases:

- ✔ **Work in progress:** This is where non-verified data is stored and exchanged internally by individual teams or disciplines such as architects, engineers, and so on.
- ✔ **Shared:** After discipline data is verified, it's shared with the wider project team for coordination and other use cases.
- ✔ **Published documentation:** After it's published, the total project team can use the coordinated and validated design output, such as information to procure from or build from.
- ✔ **Archive:** After it's built, the record information of the project history will be archived in the CDE for knowledge, regulatory, and legal requirements.

Checking and signing-off procedures allow those managing the information to transfer validated data between the work phases. During the project design, procurement, and construction phases, the CDE forms the basis of the project information model (PIM). Then when handed over to the employer/occupier, the CDE forms the basis for the asset information model (AIM).

Understanding IPD

Integrated project delivery (IPD) collaboratively brings people, systems, business structures, and practices together to harnesses the talents and insights of all participants, in order to optimize value for the client, reduce waste, and maximize efficiency through all phases of design, fabrication, and construction.

IPD is a delivery structure that aims to support interests, goals, and practices through a team-based approach to project delivery. The focus of IPD is the value created for the owner in the owner's completed asset through a Lean and collaborative approach. Traditional forms of contracts are often seen as a barrier to collaboration. However, IPD is synonymous with early cooperation and effective decision-making in the construction process.

IPD incorporates BIM and VDC processes to enable project teams to use information in an integrated environment and work efficiently using a series of guiding principles.

Trust is an essential part of IPD and encourages parties to focus on project outcomes rather than individual goals. Even though IPD promises better outcomes, outcomes can only change if the people responsible for delivering the outcomes also change. According to the AIA document "Integrated Project Delivery: A Guide," which you can find at `info.aia.org/SiteObjects/files/IPD_Guide_2007.pdf`, achieving the benefits of IPD requires that all project participants embrace nine principles:

- **Mutual respect and trust:** Behaviors are key to IPD. The project team understands the value of collaboration, and each member is committed to working as a team in the best interest of the project as a whole rather than their individual disciplines.

- **Mutual benefit and reward:** An IPD contract is founded on a mechanism that is a balance of pain and gain, where participants share both risk and reward. The contract is based on improvements to achieve goals in efficiency and support collaborative working. These contracts apportion responsibilities and benefits fairly and transparently, and have mechanisms for delivery that focus on trust and relationship.

- **Collaborative innovation and decision-making:** IPD contracts judge ideas on their merits, not on the author's role or status. BIM and VDC are key to supporting this process.

- **Early involvement of key participants:** The IPD model encourages the engagement of key project participants to allow an influx of knowledge and expertise at early stages, where informed decisions have the greatest effect.

- **Early goal definition:** A key activity of IPD is determining and developing project goals early, and all participants agree on and respect them.

- **Intensified planning:** Increased effort in planning results in increased efficiency and savings during execution. The goal is to improve the design results, shortening the more expensive construction effort.

- **Open communication:** Team performance is based on open, direct, and honest communication among all participants. A no-blame culture leads to identification and resolution of problems.

- **Appropriate technology:** Open and interoperable data exchanges based on disciplined and transparent data structures are essential to enable communication among all participants.

- **Organization and leadership:** An IPD project is in itself an organization where team members are committed to the project's goals and values.

The distinctive multiparty agreements of IPD, alongside pooled risk and reward arrangements, relational contracting based on trust, and a continuous improvement cycle, are an ideal backdrop against which to implement BIM and VDC successfully.

Sharing risk and reducing it

Over the years construction projects have become increasingly complex and challenging, both in terms of the scope of requirements, such as building services, and the need to be more altruistic in their solutions, such as energy efficiency. This ever-increasing complexity is coupled with lots more information (though often not enough exists at the start of a project) that needs to be coordinated; multiple parties that need to be unified and managed; considerable logistics that need to considered; and more onerous statutory requirements such as building codes that need to be complied with. No wonder construction is considered a risky business!

Each construction project is also unique in nature, and therefore the risk — for example, the exposure to loss — has a different profile in terms of likelihood and impact if that risk comes to fruition in the future. The good news is that you can manage construction risk through BIM, VDC, and IPD.

How so? Keep these sections in mind to manage risk.

Identify risk and impact

You can use BIM and its data richness to help identify the likely risks and impact right at the beginning of a project, in the sandbox of your computer. After the models have been developed even at a macro level, the project team members can undertake a first workshop in which they walk through the virtual construction environment to identify and recognize the potential risks. This isn't an automated process and necessitates having an experienced team systematically working its way through the model in a logical manner, such as following the schedule's work breakdown structure (WBS).

Ideally, you have a large space for the workshop with two big screens, one showing the models and the other a risk register that you populate during the workshop as you identify the potential risks and evaluate their magnitude. You need two people, one driving the model and one facilitating the risk workshop. Attendees should include representatives from all the key stakeholder groups engaged at that stage of the project and, ideally, proxy representatives from functions that aren't yet procured, such as facilities management. Use a checklist to capture criteria that the model may not be able to be answer through a comprehensive project risk register.

Manage risk

After you have a completed risk register, you need to understand the risks in more detail and understand how you can manage, or even better eliminate, them; for instance, by using pre-fabrication techniques. At this stage you can start to employ the VDC toolbox to simulate and understand the risks in more detail. In most cases mimicking all the elements identified in the risk register is impractical. Instead, you generally prioritize mitigating the big-ticket risk items, those that have been categorized as high likelihood and high impact if they land.

You can create simple 4D time simulations to better understand scenarios and look at what alternative processes or resources can be employed. This process isn't automated, and you need an experienced team to understand and analyze the model data. Then you can make a well-informed decision to reduce the risks to an acceptable level.

You repeat this process at key stages throughout the project lifecycle — for instance, prior to handover — to ensure a smooth and hassle-free transition from stage to stage.

Mitigate risk

You must consider how the form of contract will deal with risk and allocate it to the best party, rather than merely stepping the risk down through the supply chain where each party adds risk contingency. It makes good sense that, rather than each party adding a risk buffer at their level, a whole project approach to risk be taken. This is where IPD comes in.

Although many contractual options exist to help deliver IPD, in most cases all deal with risk much more effectively than traditional cost-led contracts. The incentive of risk- and reward-sharing across the entire project team ensures much more effective management of risk as opposed to the usual pass the parcel, where each pass costs the client and adds to the overall project cost.

The trinity of BIM, VDC, and IPD not only helps manage the project risk but can also help identify alternative scenarios that may give added value, such as schedule reduction and maximizing the chances of turning threat into opportunity.

Transforming Information and Communication

Information and communication technology (ICT) has transformed how the construction industry creates, validates, and shares project information beyond recognition. The industry has moved from the site office fax machine buzzing away in the corner to the 24/7 world of emails and file sharing. Project ICT has gone from being fixed, massive, and operated by an engineer with a bowtie to small and mobile (although funnily enough cellphone technology devices have evolved from big to small to big again in terms of screen size).

In the next sections you investigate the influence and impact of ICT has had upon the construction industry, and consider some importance questions that your project or business should consider and ways to solve those problems.

Feeling the influence of ICT

The impact of ICT on construction has been profound. It's reduced the lag in the decision-making process from weeks to days, from days to hours and, in some cases, to real time or automated. This, in turn, is changing the industry's working practices to ensure digital technology can be harnessed to increase productivity.

The Internet-enabled paradigm shift in communication has been both a blessing and a curse to the project process. Back in the world of analog, the various disciplines managed the transmittal process with some particular rigor in supervising each transaction. With the growth of email, this process of exchanging data became a bowl of spaghetti. The simplicity of using simple email or swapping a pen drive means the workflow becomes unmanaged and lacks validation. Neither of these methods is beneficial to consistency, coordination, and, most importantly, collaborative working.

Standard and consistent information management processes, agreed standards for project communication, and information exchanges are a must for the successful implementation of BIM. Essentially, only one standard should exist, and that's at project level (not at discipline level) with everyone following the same principles of best practice. This need is amplified when a project employs BIM alongside other data-intensive methods such as GIS integration.

The quantity of global digital data is growing all the time, from gigabytes to petabytes. Construction organizations are inundated with project data that they can gather and exploit both graphically and nongraphically. How this information can be moved about and how it can analyzed to help give meaningful insight are key challenges to any construction business in the 21st century.

Have a look at your project or business and ponder:

- ✔ Do you have an ICT strategy that's aligned with your current and future objectives?

- ✔ Are your staff being regularly upskilled to work efficiently with new project technologies?

- ✔ Do you need to retool for digital working?

- ✔ Do you have a default CDE solution for your projects with standard supporting processes?

- ✔ Your project digital data is valuable, so do you have cyber security measures in place?

- ✔ Have you checked for interoperability between your systems and software?

- ✔ Is the speed of information transfer fast enough, or do you need local caching servers?

- ✔ Can your staff access project standards and best practice online?

- ✔ Can you validate your information and can that checking be automated?

- ✔ Have you developed roles and responsibilities to support your strategy, such as an information manager?

- ✔ Have you developed a mobile strategy for data retrieval and creation on the project site?

- ✔ Do you have a project communication strategy?

Understanding the importance of ICT and its role for realizing communications between project members is crucial to winning work and project success. Getting these aspects right can be game changing.

Solving the main problems

The most common problem in any BIM journey is resistance to change. "We've always done it that way, and if there was a better way, I'd have worked it out by now, not you!" Yep, changing the status quo bias is the most everyday encounter for those leading a digital transformation in an organization. You can overcome it with the following suggestions:

- ✔ Keep the message simple and use business language as opposed to lots of BIM acronyms.

- ✔ Focus on the why and the opportunity of BIM rather than what it is.

✔ Have a clear business case and investors' report communicating investment information for change.

✔ Use metaphors that everyone understands.

✔ Talk about information management and project data. Your audience is more likely to have that "ah-ha" moment!

From the preceding tips, the perceived investment cost in BIM is where you're most likely to encounter a roadblock. The cost of new software, hardware, training, and the like is often a bitter pill to swallow for the management team that controls the purse strings. Your investment case must be well thought out and build a realistic return on investment plan with numbers rather than anecdotal evidence about how cool BIM is. Ensure those approving the plan see BIM as a business imperative (not a technology intervention) and that the cost of doing nothing would be more costly or even fatal.

After you have the investment cash, your problems are really about to begin. You need to determine what technology to buy. Refer to Chapter 6, where we discuss how to search for the right technology and what issues and needs you have to factor in. After you look and understood these needs then and only then can you put a specification together and get out into the marketplace for some retooling. Check out Chapter 17 for more on specification.

After you're up and running, another potential bump in the road is information management. To avoid problems in your fight plan, ensure that you have established standards for information management with a BIM execution plan created for each project and a CDE in place. Head to Chapter 8 where we explain the ins and outs of an execution plan and why having one is important.

You'll encounter additional problems and challenges, like in any change program. Don't face them alone; set up a steering group to discuss and remedy any issues. Make sure you take something away from your early pilot projects, share the lessons, and don't make the same mistake twice.

Presenting All the Incentives of BIM

As an intelligent model-based design process, BIM can add value across the entire lifecycle of a project when done well. Imagine how your project would perform if you always had perfect, coordinated, and validated information that everyone could access and share. That in itself should be an abundant incentive, but you want more!

Better stakeholder understanding and buy-in through your 3D and non-graphic data can lead to the following:

- **Improved understanding:** Everyone can better grasp the design intent and potential business outcomes through 3D visualization and simulation. Visualization and lifecycle solution testing should commence as early as the preconstruction stage. Visualization of the solution in 3D can really reduce overall the time schedule, especially if your project has large amounts of stakeholder engagement. Additionally, the number of requests for information is likely to reduce dramatically. Ultimately, BIM is about actually giving your customers the buildings they want and can efficiently operate and maintain.

- **Lower costs:** Open, shareable information means better understanding, which means less risk, and less risk means less contingency. Contractors and subcontractors comprehend their work scope better at the pricing stage, and through building better before built they can look at optimization of their time schedule and associated preliminary costs such as construction plant machinery.

- **Safety:** BIM can help support a safe-by-design agenda where all participants can interrogate and validate the model for both buildability and operational safety. Logistical scenario planning and visual method statements can easily be generated. However, it's again back to the first point that BIM better helps understanding and communication especially for inductions and workforce toolbox talks.

- **Energy and carbon:** BIM can help create more altruistic solutions that are cheaper for the customer to operate. Through early model analysis and reviewing various options, you can take a whole-life approach to the design, ensuring the building will be energy efficient and have a smaller carbon burden. The cost of building operation (energy and maintenance) is much more significant than the capital expenditure — for example, the cost of designing and constructing the building — which is a massive incentive for the client.

- **Better coordination of project information:** No more clashes! Actually, the big benefit is that BIM coordination sessions get everyone together and help create a less adversarial project culture. Coordinated project information means the project team can validate digital information so that you can make the right decisions and make them more rapidly.

- **Improved productivity:** Retrieval of information from a CDE alongside more Lean and BIM supported work flows improves productivity and the exchange of project data more collaboratively with little waste. Here's an important point to note: BIM in isolation won't boost productivity significantly; you need to couple it with a modernization agenda, especially VDC.

- ✔ **Reduced rework:** Having real-time access to coordinated information, especially using mobile devices, to automated field technologies and work flows such as setting out direct from the model are all helping to reduce the extent of rework that prevails in the sector.

- ✔ **Increased speed of delivery:** 4D simulation and Lean working help ensure that the schedule is optimized at all stages of the plan of works for a project. Using object libraries and maximizing the design for manu-facture and assembly opportunities improves the speed of delivery.

- ✔ **Better outcomes:** Ultimately, the prize is giving the customer the asset he wants with a soft landing and performing as it was intended every time. Early engagement of the operational teams in the process seeks to reduce the hidden costs of adapting completed spaces to suit specific end-user needs, which is an important part of the BIM agenda. BIM gives greater program and outcome certainty — you've tested it in the safety of the model.

- ✔ **Improved operational management:** This includes digital working meth-odology in the design office or on-the-job site, from digital quantity take-off to setting out from the model (these benefits are unlimited). From automatically populating the computer-assisted facilities management systems to operation and maintenance data on a mobile device, BIM has a big part to play beyond delivery. The AIM is a natural progression from the PIM. Containing product and warranty data, these operational digital models greatly assist in the smooth management of the asset.

Part II
Creating the Foundations for BIM

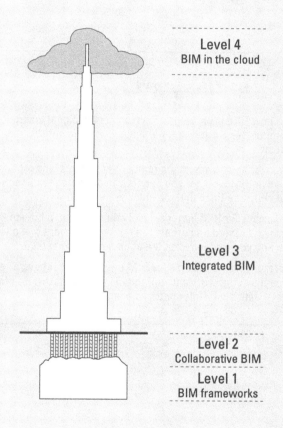

Level 4
BIM in the cloud

Level 3
Integrated BIM

Level 2
Collaborative BIM

Level 1
BIM frameworks

Head to www.dummies.com/cheatsheet/bim for a handy Cheat Sheet that you can refer to whenever you need to know about BIM.

In this part . . .

- ✔ See how BIM is part of the development in digital technology that provides a catalyst for change within the construction industry.

- ✔ Discover how to select the right projects, tools, and technology to make BIM work for you and how to encourage colleagues and convince them that the industry needs BIM.

- ✔ Establish your BIM maturity level and understand how to keep pushing forward and create the right environment that gives you and your organization the edge on the competition.

- ✔ Start to deliver collaborative BIM processes that have a real impact on your work and understand how to make your life easier with digital information management.

Chapter 6

Looking at the Digital World Around You

In This Chapter

▶ Explaining that technology is a catalyst for change

▶ Identifying your needs and communicating your requirements

▶ Selecting BIM tools without breaking the bank

▶ Recognizing that BIM will shake things up, but keeping the good stuff

You can't buy BIM in a box. BIM is a process, and to be successful it requires people and technology to work together. The construction industry is full of information technology. Balancing acquiring new skills required for CAD software or energy analysis and building codes programs on top of day-to-day project communications that fill your email inbox with attachments and drawings is difficult. So the idea of implementing a brand-new process that relies on new technology can seem pretty daunting. The key is to implement BIM by making lots of gradual improvements in the technology you use. This chapter shows you how easy that can be.

This chapter examines the digital world around you and outlines what lessons you can apply to the way you implement BIM in your office or workplace. In this chapter, you see that technology is often the driver behind dramatic change and discover how to convince your boss or your colleagues that the construction industry is no different. By looking at other industries, you glean some great examples of disruptive technology – the kind that really shakes things up. You also find out whether the technology you already use can become part of BIM or whether it needs to evolve.

Targeting Technology

Unless you're reading this chapter after a long walk in the mountains, you can probably see a lot of technology right in front of you. From your smartphone or MP3 player to your digital watch, smartwatch, or laptop, you surround yourself with increasingly linked devices. Perhaps you're reading this chapter on a tablet or other device as an e-reader. Most parts of today have gone digital. In fact, it's difficult to think of many aspects that are unaffected by the digitized, connected world.

BIM is part of that connected world finally impacting construction. The following sections look at why technology changes mean you need to do your research when you're selecting technology to adopt. We also show you ways to encourage your colleagues to try something new.

Understanding why technology evolves

Just think about all the various ways that you use technology on a daily basis. You probably use a smartphone or tablet to check your email and social media accounts as much as you do a desktop computer or laptop, or access the Internet via free Wi-Fi in cafes and airports. Most cars have very complex on-board computers too. You may have a fitness-tracking device, play games on a console, or stream video to your TV to catch up on your favorite shows or films.

A better question to ask may be: "What do you do that doesn't involve digital technology in some way?"

Technology has often replaced an equivalent physical format like paper books, vinyl records, 8 track tapes, or magnetic tape cassettes by developing a digital version. The evolution often makes it easier to replace analog equipment like record players or VCRs designed to read one format with one device that does everything. Clearly, the change in format provides lots of obvious benefits, like how portable the item becomes — being able to go running with all your music on your device or not having to take lots of paperback books to the beach. But what's actually important about digitization isn't just the obvious physical format change, as useful as portability may be.

The reason for all these added benefits is that digital technology provides a place to add data into the mix. As well as the actual product like the latest album, film, or book at your fingertips, embedded within the digital object are thousands of packets of information (When was the song released? How long is each movie scene? What is the next book in the series?). This additional content gets transferred with the object, no matter how many times it's replicated. You can call this additional content the information or data about the object. This is information modeling; in your case you're modeling information about the process of building something, whether it's infrastructure, civil engineering, a skyscraper, or a city plan.

Doing your research when selecting BIM technology

Going out to choose new technology can be an intimidating process. The most important thing is to understand your needs. Determine what problems with your current processes you're trying to solve and ensure that the technology, products, and services that you consider are real solutions.

The key to making BIM work for you is that its implementation should add value and improve efficiency right across your operation. This section gives you some quick tips to make sure you have the best chance of selecting the most appropriate tools and platforms for your business. Vendors use the term *platform* to describe their BIM products and that's what software is: a platform for building on, taking off from, or creating something new.

With new technology, you have a lot of decisions to make. To ensure that any new systems you implement are useful, consider these key tips:

- ✔ **Research.** Investigate each possible alternative before committing to anything. Use vendor websites to find out the key facts that may affect you, such as cost, training options, or file formats. Make sure you differentiate between the promises of sales brochures and the details that directly apply to your business and your office.

- ✔ **Ask the people you trust.** Everyone is in the same boat. Rest assured that every office and every business (no matter the size) is at various stages of BIM implementation and must make its own decisions. Benefit from the knowledge and experience of peers and contacts across the industry by asking them for advice. Certain information brings a commercial advantage, so some people may be reluctant to share confidential plans, but you'll find that most are vocal with their opinions.

- ✔ **Compare features against cost.** Focus on the essentials. Remember that software with the most bells and whistles isn't necessarily the best option. For your needs, you may only want simple functionality or a cheaper, very specific product that offers an ideal solution.

- ✔ **See whether you can demo.** Ask whether you can test-drive products, services, and other technology before you buy. Ensure that you use the demo time efficiently, carrying out real tasks and recording metrics like time and ease of use.

- ✔ **Be prepared to change later.** Don't constrain yourself to one option too early. Think about what products you may want to link and integrate with other systems and technology, especially existing processes that are important to you.

Using BIM technology to your advantage

As with any new technology, think about the benefits it can provide in replacement of existing processes. For example, compare how you currently determine the energy efficiency of a building with the various tools, software, methods, and guidance available. Look at the many CAD modeling programs that offer in-built tools for determining environmental performance, often to demonstrate meeting regulatory requirements. As part of BIM, you can combine this data with briefing and performance specification, location-based factors, environmental consultancy, and product vendor values to refine energy-efficient design of buildings.

Any technology used as part of BIM implementation should provide the following key benefits:

- ✔ **Find additional convenience or efficiency.** If you're struggling to locate information when you're searching for it or your current technology is slowing down your working processes, look at alternative tools.

- ✔ **Do everything faster.** You live in a world obsessed with having the answer yesterday. Instead of complex calculations and measurements, you can run reports and take-offs from CAD and specification software faster than ever before and make automated processes more efficient, tailored to your workflows.

- ✔ **Access knowledge and information.** The Internet and digital publishing has opened up a wealth of linked and updated information such as building codes, manufacturer and vendor data, specification guidance, and best-practice method reports. Alongside proprietary information, you can build your own digital library of BIM project data. Find opportunities for using digital library content, such as learning from an error or achievement in a previous project and adding that knowledge back into the office database.

- ✔ **Combine data.** A critical aspect of BIM success is the ability to bring together multiple strands of data into one report. This may be for the purposes of clash detection; for example, working out where pipework collides with a concrete wall.

- ✔ **Increase communication between team members.** BIM technology should make cooperative work easier — for example, coordinating the delivery of landscape design based on information from civil, structural, architectural, services, environmental, and landscape design teams.

Encouraging people to use something new

In order to ensure that BIM has the best chance of success, you need to give people the best chance of understanding what it can do for them, at their desks, every single day. You need to prepare positive and effective training that will support users, not only to understand BIM as a theory, but through the inevitable changes to working practice and processes. In the next section, we discuss change management. Chapter 14 looks at change and implementation in more detail.

Put yourself in the position of someone handed a new electronic device without warning. You're told this device will make work easier, make your day more efficient, and help you to communicate with project teams. However, you've received no instructions and no training. You're not even sure how to turn it on. What would happen to the device? Would you try to work things out for yourself? Would you believe what you were being told about its potential? That's what BIM can feel like for busy people who keep hearing about it but aren't kept in the loop.

Identifying the Need for Change

If we asked you to name the processes and departments that make your business less efficient, you'd probably be able to think of some very quickly. Ask some colleagues what they consider to be the problem, and you may get a totally different set of answers. What you think of as outdated and obviously disorganized procedures can make complete sense to other people, especially if they've worked in that way for many years.

To some of those in the various roles impacted by BIM implementation, the idea of changing some processes will seem impossible or highly disruptive. This section demonstrates ways that you can identify the need for change and see where BIM implementation will improve the efficiency and quality of your organization.

Before you can see what effect BIM will have on your business processes, understand where change is needed most. After you have that information, you can more easily and accurately plan BIM implementation in terms of your existing systems and especially the day-to-day users. There are excellent examples of best practice BIM work flows; for example, the UK Cabinet Office BIM process mapping diagram, which can be found at `www.bimtaskgroup.org/wp-content/uploads/2013/05/LG-BIM-model-process-map.pdf`.

To detect the need for change, we suggest that your company runs an internal systems survey with all staff. Make the survey anonymous so that everyone feels free to comment openly about problems and inefficiencies. Ask where bottlenecks and pinch-points exist in their daily jobs. Remember to also ask users what they think works really well.

Be aware that changing existing processes can have a huge impact on people and be a very stressful time for some staff. Consider social, psychological, emotional, and even physical effects as part of a full change management plan. Chapter 14 provides more information on change management.

Consider costs now and tomorrow. Don't just look at the upfront cost when considering your options, but think about the various additional costs the software, technology, or service will generate. Remember that a cost may exist for implementation and setup, along with annual running costs. Bear in mind that the initial price may not be guaranteed to be the same in Year Two or when the next version or update is released. Ask about the cost of training and fixing problems too; many vendors provide software support free of charge.

Keep the long term in mind. Focus on the essential processes and functions that will impact you from day one, because you need to measure return on investment (ROI), but don't lose sight of potential development, both in technology terms and in your own business. If the software or services you consider have additional features you won't use, take the time to think about why that is and how you may add them to your future processes.

Defining Your Requirements

Figuring out what you need as opposed to what other colleagues need or what the business as a whole needs can be a tricky part of technology implementation, because the answers you receive can be very different. Defining your requirements ensures that BIM implementation always focuses on solving real issues and providing added benefits specific to your business and the work you do.

In the earlier section "Identifying the need for change" we suggest that you run a survey on internal systems in your business. Take the information you gain from this survey and interpret it. Pay particular attention to processes that are breaking down or causing delays. You don't have to communicate your findings to others, but you may find benefits further down the line in your implementation strategy if you've let the respondents know that you've listened to them.

Everyone has to go through a number of aspects of BIM implementation. The following sections help you to plan for these common processes, including a technology and systems review and justification of investment required.

Noticing common issues and themes

You can position BIM implementation as a direct response to the concerns and problems flagged by your colleagues, but start with the obvious things. There are likely to be common themes about a particular program, office equipment, or Internet speed that can be relatively easy to solve but can provide huge benefits and cost savings.

These different concerns need various approaches to finding solutions:

- ✔ **Systems that are obsolete or obsolescent:** Some of your systems will eventually fall into obsolescence, or become obsolete.

    - *Obsolescent* describes something passing out of usefulness. Some vendors use *planned obsolescence* to encourage buyers to upgrade to the next product, including some technology manufacturers.

    - *Obsolete* means something no longer in common use, generally because it has been replaced by something better.

    This issue can be relatively simple to resolve, because new versions of systems and software may be available that your organization has never considered or deemed too expensive. Investing in these advanced releases can have a significant impact on productivity. Look for free updates to software or support fixes that will upgrade system performance.

- ✔ **Frustration at time wasted:** This concern could have multiple causes not related to technology at all, but often they're simple factors such as Internet speed or delayed system computing power. Contact the relevant system providers and vendors to see whether you can make performance gains by changing configurations or settings.

- ✔ **Duplication of effort or having to rework mistakes:** This issue is the very kind that BIM implementation aims to solve. Aim to change internal processes to coordinate information more efficiently and avoid repeated errors from project to project.

- ✔ **No access to project information:** This is a traditional problem in many offices. Finding critical data about the performance of a system or vendor data about the supply of a product can be difficult, especially if other teams are working on the project. High-quality BIM objects carry hundreds of properties for this kind of data. *Properties* refer to characteristics that can be assigned to objects to reflect specific information — for example, technical data or functions for designing, calculating, or constructing the object. Note that the terms *parameters* and *attributes* are also used in BIM platforms and construction information. Collectively, these bits of information form a *property set.*

Because objects are information in their own right, this additional content is called *metadata,* which is data used for the description and management of documents or other containers of information or the data content. Literally meaning *after data,* this term just means to describe a level of abstraction from the information held in each object.

A good way to review your requirements is to run through an imaginary perfect process and record it. Take a large sheet of paper and in one color draw out each stage of how things should work together and approximately how long tasks should take. Now compare this with your last project by adding real timescales and actual project tasks to the diagram in a different color. This exercise can provide a clear visual representation of where problems occur.

Investigating how BIM can help your team

You should be able to explain in your own words what's wrong with the technology and processes your business uses on each project prior to BIM implementation. Many of the issues are obvious and BIM can create change in lots of ways directly in your office. However, the ideal is to consider BIM across the entire project timeline.

Here's a list of the benefits of BIM that will underpin all future project work in your organization:

- ✔ **Prioritizing health and safety:** Keep site safety in mind constantly. BIM implementation must reduce the number of construction site fatalities by improving site planning and operation. BIM should encourage safer construction and fabrication methods.

- ✔ **Reducing the need for changes:** Build everything digitally first, long before ground is broken on-site. Many BIM platforms provide clash detection functionality to highlight building issues, especially between coordinated information from multiple teams.

- ✔ **Improving design and construction quality:** Don't believe anyone who says that BIM removes the element of design from the built environment.

Use the power of the BIM platforms and related tools to push the boundaries of what's possible. Overall, you can improve construction quality because BIM best practice should resolve every detail. Consider everything from a contractor's point of view.

✔ **Prefabricating:** After you design each element and define quantities, you can consider prefabrication far earlier in the process than ever before, such as pre-cast concrete or pre-welded steel. Think about the opportunities for off-site manufacturing and design for manufacture and assembly (DfMA), speed of construction, site safety, and build quality.

Starting with what you already have

Everyone has to start somewhere. Some of your existing systems and processes will survive the advent of BIM-focused procedures, but you'll need to replace others or evolve them to meet the demands of a progressively more informed industry. To determine the quality of what currently exists, you can run a series of investigations.

An understanding of your technology and internal systems is important, but never lose sight of the bigger picture. BIM isn't about software, and its cost shouldn't be an obstacle to implementing BIM processes and improvements. The research you do now ensures that you're investing in the right software and platforms, which can make the wider change management process far more successful.

To define and analyze your processes and systems:

✔ Identify hardware and record the specification, age, and performance of the machines.

✔ Review software, including its version number, update process, and its usage if possible.

✔ Recognize up-front costs and the cost of licenses or subscriptions required.

✔ Understand infrastructure, including Internet service provision and the server arrangements your company data is built and stored on. Consider that external companies and data centers may provide storage.

✔ Categorize basic devices such as phone systems, which can still be integral to even the most online, agile, and efficient of modern offices. Consider finding an alternative to the fax machine.

✔ Include off-site tools such as company laptops or mobile phones that form part of your wider network of connected devices and business systems.

You may feel confident conducting this type of systems research yourself, or you may be part of a company where a business systems team collects and manages this data on a regular basis, saving you some work. If not, you can employ consulting companies to do various levels of systems analysis for a fee.

Investing to make long-term savings

You can understand that some investment is necessary for any business development. Often an upfront cost, investment in BIM can also include time, resources, ongoing costs, and other overheads. For any outgoing budget for systems or resources, you want to see a return on investment as soon as possible. Now that a great deal of BIM case studies and reports are displaying benefits and efficiencies, you may feel additional pressure to demonstrate savings and profits early. However, realize that BIM is a long-term investment. You may be able to show that BIM is an immediate advantage to your business but the real benefits are more likely ongoing and gradual. You may need to convince senior finance staff that the time and money required for BIM will affect processes they're not necessarily ever likely to be part of on the ground. By making these process changes on the shop floor, you'll see larger efficiencies.

You need to justify the cost of BIM implementation to the business. Some companies consider passing on the cost of their BIM implementation to their clients or to increase fees related to specific projects that are using BIM processes. Think carefully about how doing so affects your commercial advantage; for example, when tendering against other businesses. Put yourself in your customer's shoes — would you think it was fair that investment in making a profitable service better was directly passed on to you? You may write off the cost as part of business income development. Consider the options for capital allowances in your region. Grant assistance programs may also be available to contribute to the costs of BIM implementation.

Involving everyone in BIM

Although focusing on the impact of BIM on your personal situation and everyday processes is important, keeping every member of the project team, from client to end users, in mind when developing BIM strategies and implementation plans is beneficial. Here are a couple of short sections that contain tips for engaging everyone at all levels of your organization.

Be clear how management will measure a return on BIM investment

BIM savings don't necessarily equate to instant business profit. You may see efficiencies and savings that are practical and method-based and so not immediately economic. You can attempt to calculate return on investment from IT projects and system upgrades, but doing so is notoriously difficult. The fundamental return on investment calculation from a financial point of view may not suggest cost-effectiveness.

We're fortunate enough to have worked with some forward-thinking senior colleagues who understand that the traditional way of looking at IT systems cost as contrasting with process management and operational cost is now out of date, and so were prepared to finance the future of the company. Because everyone uses technology every day, equipment and process are so indistinguishable as to be impossible to separate in any meaningful way. You may have experience of more cynical or bottom-line obsessed managers who still want you to make this distinction clear.

Invest in people and skills as much as technology

A good general rule is that you aim to spend as much on training as you do on the BIM platform or hardware equipment. That may not be a literal cost of paid software training; instead you may provide the equivalent time for practice outside of project work. You can propose the following kinds of sessions:

- Overall IT business systems investment and how it fits into BIM implementation. Try to get executive staff members to present their vision.

- Introductory and specific product training on functionality and benefits of BIM platforms and tools, as close to project usage as possible.

- Linkage between new software and other systems you use, including discussion of any nontechnological processes affected.

Demonstrate intangible benefits to stakeholders

Although return on financial investment is always how business costs are measured, it doesn't always tell the whole story. People are the most important barometer when assessing how successful change has been. They're far better than anything numbers or spreadsheets can tell your senior management and company stakeholders, especially in the first few years. By showing how BIM will benefit staff and generate new efficiencies, you'll make much better progress toward financial return. Always deliver a regular update to sponsors that

- ✔ Indicates the more intangible returns on investment that you see resulting from replacing and upgrading systems and software platforms.

- ✔ Provides real examples of improvements in inter-team communication, new creative activities, and, especially, enhanced customer service or great client feedback.

- ✔ Focuses on facts. You're far more likely to get engagement from senior stakeholders if you keep concise. Don't send them 50-page documents or forward unedited email streams.

You can make sure that everyone is set up to help one another by nominating super-users. *Super-users* are the most experienced users of the software in question, but should also have the communication skills to become first-line support for colleagues. You can also provide support through accessible frequently asked questions pages published in a central location, such as your company's intranet. Set up regular group updates where users can shape cooperative working.

Developing this sense of community can be half the battle. Improve team morale and office atmosphere through BIM's more communicative processes, and the increased productivity and job satisfaction easily outweighs the initial investment in BIM.

Selecting BIM Tools

BIM isn't software and certainly not just 3D CAD. The idea that you can purchase BIM proliferated because software vendors were the first to push the parametric object modeling capabilities of their programs. Most people are aware that BIM is more than just software and technology. However, some areas of the industry and people in certain locations around the world still have that perception.

That said, you have to make decisions about BIM tools and platforms at some point. These sections focus on four key things to help you.

You may work in an office that has a dedicated team for sourcing and procuring software tools for the business. You may be concerned that department isn't listening to your advice but is prioritizing basic IT considerations. BIM implementation affects more than just technology, so work with senior colleagues in that team and elsewhere in the organization to ensure that the systems team carries out complete requirements analysis and that you're involved in proposing, testing, and discussing the software options. Don't let detail-level considerations like ease of deployment, update processes, or administrative permissions become the focus of discussion when management must make fundamental choices that will impact business processes.

With the right information and approach, you can be the conduit between executive-level staff, management teams that have responsibility for different areas of the business, and daily users of the software. Executive staff members won't thank you for blinding them with science, so jargon and detailed description of IT infrastructure or software functionality aren't useful. Always describe the process that new software enables or that requires a solution. Equally, front-line users aren't going to welcome phrases from business cases and financial proposals; they want to see real, effective change occurring.

Avoiding the hype

New and shiny ideas that take hold in the industry can heap pressure on companies to stay relevant and appear ahead of the curve. The requirement to demonstrate sustainability led to a huge amount of *green-wash,* inflated credentials about how sustainable companies were. Green-wash could be found across the built environment sector, from contractors indicating how much of their material was recycled to product manufacturers demonstrating sustainable sourcing of timber or responsible water use. You've probably noticed exactly the same thing happening with BIM and this type of over-blown hype has been dubbed *BIM-wash,* the unrealistic statements of how BIM-capable a company is.

BIM-wash and posturing about BIM competency has a negative impact on trust and relationships across the design team. More and more clients and employers are asking for BIM at its various levels but not understanding what they're really asking for. Tenderers such as designers, contractors, and consultants are claiming to be able to deliver against the requests to win the job — creating a maddening loop of vague requirements and subsequent failure to deliver. Based on this confusion, many of the most popular BIM tools and software platforms claim to be the ultimate BIM solution or to provide a dedicated answer for your particular discipline. The topics you're trying to solve with BIM tools may include

- ✔ Existing slow or poor systems (for example, file storage or email document exchange, full memory, making the most of cloud services)

- ✔ Lack of communication (for example, between members of a team, between multiple teams or between multiple companies)

- ✔ Need for additional data and metadata (for example, searching for who added the information and when)

- ✔ Issues with collaborative working (for example, wanting to combine multiple information sets into one connected, federated model, and running clash detection reports)

- ✔ Ability to update information as required (for example, changing a property like a door size globally across all views and documents)

- ✔ Quality and speed of visualization of the project model (for example, producing high-quality, realistic rendering of the project)

Finding out what's best for your organization

People give opinions and reviews of BIM tools, platforms, and software systems all the time, especially on industry blogs and in magazines and journals. Keep an eye out for objective reviews rather than paid advertisement features and gain as much information as possible before making your selections. Read through all you can to determine which ones fit your organization's needs the best.

As you can see, BIM tools have some obvious and not-so-obvious benefits. You can organize a number of ways to compare and contrast functionality between software vendors.

If you arrange the following types of sessions, you can also increase user engagement and generate enthusiasm. Here are some good ideas for how to increase awareness and shared knowledge of BIM software and supporting systems:

- ✔ **Have existing users in your organization demonstrate some software.** Doing so can be very informal and inexpensive. You can plan a Q&A session as part of this demonstration and ask colleagues to submit questions in advance, allowing you to filter and consolidate concerns or queries.

- ✔ **Invite an external vendor to provide a continuing professional development (CPD) session over lunchtime.** Sometimes these sessions can turn quickly into sales presentations, so clarify your agenda with the sales representative and say that you'll discuss costs or licensing options outside of the group meeting.

- ✔ **Arrange a site visit to another office or company that uses the software.** Paul has previously arranged one of these sessions and he found it much easier to organize with a company that works in a different sector or discipline to avoid any competition or privacy issues getting in the way of what's just an open investigation. If the company seems reluctant, offer to sign nondisclosure agreements about anything that colleagues see or discuss. This is a great strategy when you're gradually rolling out BIM applications between multiple offices. After the first implementation, you always have a very relevant case study to learn from that can produce excellent cross-office communication and team-building.

✔ **Book meeting time for internal review and brainstorming sessions.** The collective wisdom of the group provides good insight into the implementation, beyond what you've already considered. Others may be able to share personal experience or suggest software and tools that you haven't encountered.

Discovering free tools

Alongside a variety of licensed proprietary software, many BIM platform vendors are part of an organization called Open BIM (an initiative of buildingSMART) or the BIM Vendor Technology Alliance, so they'll provide free tools too. You can especially find free software for viewing (not editing) model content. Some of these tools are industry specific, whereas others are more generic and made for non-designers:

✔ Open BIM viewers, designed to allow navigation of a range of open file formats including Industry Foundation Classes (IFC), BIM Collaboration Format (BCF), and gbXML

✔ Clash detection model-checking software, interrogating model information, and change control reviewer apps for managing revisions

✔ Proprietary viewers to review architectural or MEP content designed using specific BIM platforms

✔ Viewers with the ability to add comments, record screenshots, and embed model information into presentations

✔ Mobile device model readers to explore BIM visualizations as walkthroughs on tablets and smartphones, such as the augmented reality (AR) model viewer

✔ Virtual process engineering viewers

You can find some software and services offered on a freemium basis. The *freemium* model limits access to certain aspects or sets restrictions on usage in the free version, but you can unlock additional features for a premium charge. Some of the free versions of software offered using this business model can often be suitable for certain sizes of practice or certain commercial uses. You can use freemium models as a good way of trying out some services without committing to the cost of the full application, particularly for online services like virtual conference and meeting tools or file storage and database providers.

Data theft is a bigger commercial risk than it's ever been, and security of your data has become a paramount concern. In any CPD sessions, demonstrations or trial versions, make sure that you're maintaining standard data protection and data security policies throughout; for example, not using live customer data or project content. As the industry moves into an increasingly mobile and connected working environment, you need to ensure that smartphones, tablets, and devices are secured and don't provide an opportunity for someone to steal personal or confidential data. During the investigation phase, colleagues may want to download applications and trial versions of software. Centralize this process to maintain anti-virus and anti-spyware protocols.

Realizing that BIM will shake up things

You may hear BIM termed a *disruptive technology,* meaning that it has the potential to dramatically affect the construction industry by upsetting traditional processes and wiping out some obsolescent systems altogether. You can only ensure that your business is prepared for what you already know:

- You probably use an electronic key fob to open your car but, like the vast majority of people, use a metal key to open your front door. Smart homes will begin to resemble the on-board electronic systems of cars more and more as the Internet of Things and embedded sensor technology become commonplace.

- Product manufacturers will continue to develop new materials to provide greater efficiency in manufacturing and high performance or energy benefits in use, such as aerodynamic paint finishes, ultra-thin but efficient insulation, or very lightweight but immensely strong cladding panels.

- 3D printing has moved from a craft peculiarity to a potential solution for the mass production of housing or infrastructure. The advancement of robotic engineering will automate many manual and labor-intensive processes across the industry. See Chapter 20 for more information on the future of BIM technology.

Don't expect to do the entirety of BIM implementation yourself. You may feel incapable of doing certain facets of BIM implementation and that's perfectly reasonable. Maintain a positive and enthusiastic overview of the entire process, no matter what hurdles you encounter. Some things will fall into place very quickly, but you may feel like you're facing a brick wall with other aspects. Demonstrating the benefits of BIM while subtly indicating that existing processes are in need of drastic review can be difficult.

Recognize that BIM is a disruptor with the power to completely transform construction and the built environment and you're on your way! In fact, you're just getting started.

Taking a lesson from the automotive industry

Cars are a great example of keeping the user informed. If your tire pressure goes down, the car can warn you, or can even re-inflate the tire to keep the car at maximum efficiency. Some cars are hugely advanced; in Chapter 3 we talk about Formula 1 or NASCAR pit lanes using information modeling to understand the minute physical and electronic improvements that can equate to huge performance gains out on the track. Other feedback is fundamental, either because it's useful like your fuel gauge or because it's health and safety like beeps that tell you that you haven't fastened your seat belt.

So can you imagine a world where everyone drove around in Model T Fords? It wouldn't make sense — the automotive industry thrives or fails on its ability to revolutionize people's perception of cars, help drivers save money, and appeal to their current tastes, such as a more sustainable, energy-efficient future. From hybrid vehicles to driverless cars, vehicle makers constantly innovate. Of course, some people still want to preserve and maintain classic vintage automobiles like a Model T, in the same way that some people are vinyl music aficionados, but they wouldn't expect everyone to join them or suggest that their passion represents the modern world. Construction has used very traditional processes for a very long time indeed.

Chapter 7

Preparing the Foundations for BIM

As the saying goes, "Success is all in the preparation." Getting the fundamentals right before starting your BIM journey will stand you in good stead and put you in front of the competition. But before you start, ask yourself a couple of questions. What's wrong with what you're doing already? What are the practical things you have to do to implement BIM processes? Just what type of software, hardware, and projects do you need to focus on?

In this chapter, we explain the key components of BIM, starting with the bedrock, moving to the foundations, and finishing with the superstructure. We also look at what you need to consider to get yourself fighting fit and BIM ready.

Progressing through the Levels of Maturity

Simply buying and installing a few copies of a 3D BIM package and saying that you're BIM ready isn't going to cut the mustard. *Maturity levels* conceptualize growing levels of capability and outcomes in the model environment. The industry needs a useful way in which companies, organizations, teams, and individuals can theorize growing levels of capability and outcomes in the model environment.

Maturity levels together with their allied processes and tools help establish a benchmark for comparison and aid understanding as to what level an organization is at with its BIM implementation. Furthermore they prove clear competence levels that are expected, together with the supporting standards and their relationships to each other and how the project team can apply them to projects and contracts in the industry. They're useful as a way of setting a clear definition as to what's required for an organization to deem itself BIM compliant; maturity levels show the industry the adoption process as the next steps in the journey that's taking the industry from the drawing board, to the computer, and then firmly into the digital age.

As Chapter 9 discusses, the UK government has been clear that it's explicitly targeted Level 2 BIM in the UK maturity model, affectionately known as the *BIM wedge.*

Maturity models establish a datum that you can measure against and that help an organization understand where it is now and where it needs to be. They're a simple way of communicating expectation. However, you should always use maturity models in conjunction with explicit employers' information requirements (EIR) setting out a project's data needs. (Refer to Chapter 8 for more about EIR.)

Although a number of maturity models are relevant to the construction industry, only a couple makes the specific claim that they measure BIM-specific maturity:

- ✔ **The NBIM I-Capability Maturity Model (CMM):** It forms part of the US National BIM Standard developed for users to evaluate their business practices along a continuum or spectrum of desired technical level functionality. The idea is that you can use the CMM tool as a way to plot your current location and plan ahead for your goals for future aspirations. The CMM isn't new and has been used within the software industry for some time. However, unlike the CMM used within software, the NBIM CMM addresses supply-chain issues and its maturity levels take into account different project stages. The CMM is available in two forms:

  - A static Microsoft Excel workbook consisting of three worksheets

  - An interactive version that consists of a multi-tab Excel workbook

  Download the CMM at `www.nationalbimstandard.org/nbims-us-v2/doc/Interactive_BIM_Capability_Maturity_Model_v_2_0_NBIMS.xls`.

- ✔ **Indiana University's BIM Proficiency Matrix:** This matrix is an evaluation tool used to assess the proficiency of a respondent's skill at working in a BIM environment. It's used as one of many selection criteria for a given project and communicates the owner's intent to the design team members what the BIM objectives are. It covers categories such as physical accuracy of model, integrated project delivery (IPD) methodology, construction data, and facility management (FM) data richness. Respondents are awarded points and categorized as either Working toward BIM, Certified BIM, Silver, Gold, or Ideal. The matrix is a dynamic tool that will adjust as the industry matures.

You can download the IU BIM Proficiency Matrix at `www.iu.edu/ ~vpcpf/consultant-contractor/standards/bim-standards. shtml`.

Providing the Right Structure

Like any good built structure that's to stand the test of time and have longevity, your approach to BIM must be built with a firm bedrock, foundations, and superstructure. In the following sections we explain what components go into making your structure firm.

Figure 7-1 shows the big-picture glance of the following four levels of BIM. As you can see, the only way you can reach for the skies with BIM is to ensure that you have the fundamental BIM frameworks, reinforced collaborative processes, and the development of gradual integration and exchange of data in your project. Eventually, BIM will rely on cloud solutions, but there is still a long way to go.

Preparing the bedrock: Level 1

The right setting should be primed for BIM, and clear, unified goals for data use and outcomes are a smart starting point. Collaboration should always be meaningful; however, don't collaborate for the sake of collaboration. BIM doesn't eliminate the need for good design management. If anything, it amplifies the need for it.

The bedrock should comprise the following:

- ✔ **A framework for collaborative working:** This framework should be within your organization and detail how you'll work well with others, ensuring that you have a clear purpose. (This framework can answer the "why" question.)

- ✔ **A common methodology:** Establish an approach for managing the production, distribution, and quality of construction information in a common data environment (CDE) so that everyone can access the same data and so that no removable media storage devices, such as memory sticks, can be removed from your computer. Governance of the CDE is also essential. Proactive quantity assurance is better than reactive repairs.

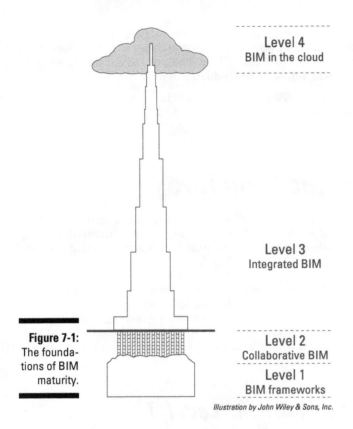

Illustration by John Wiley & Sons, Inc.

Figure 7-1:
The foundations of BIM maturity.

BIM must be founded on a solid bedrock. Getting Level 1 embedded in your organization should be the first horizon, and these factors apply even if you aren't doing BIM.

✔ **The right procurement route:** This helps set the right environment for BIM. Essentially, the more collaborative the contract, such as NEC, the better the BIM opportunity. This isn't just the client to principal contractor but also to any subcontractors. The contract can help set the scene for playing nicely. Evidence suggests that you achieve better efficiencies when BIM is part of the contractual arrangement.

Setting the foundations: Level 2

The second horizon involves the base in which you build your BIM foundation. Level 2 is a staging post in the BIM maturity journey, and you'll eventually move from collaborative data exchanges to one that's integrated, real time and cloud based.

In order to create this foundation, you first need to have processes for information management during the capital delivery and operational stages of the project lifecycle. Start with the end in mind, and understand what data you'll need to successfully operate and maintain the asset.

Key components include the following:

- **Asset information requirements (AIR):** They define the information that is required for an asset information model (AIM).

- **BIM execution plan (BEP):** Suppliers prepare this pre- and post-contract document, setting out a structured, consistent process for how the project will be carried out, including common terminology for job titles, descriptions, responsibilities, and processes.

- **Master information delivery plan (MIDP):** This document is used to manage the delivery of information during a project.

- **Organizational information requirements (OIR):** These describe what information is required by an organization for asset management systems and other organizational functions.

Also consider capability assessment. In the UK a standardized document, PAS 91, examines the supply chain's ability to deliver a Level 2 project with particular regard to information management compliance relative to PAS 1192-2.

You also need a suitable means for transporting your data across the project lifecycle and facilitating a means of validated information exchange. Remember to look at how your data needs to go from a project information model (PIM) to an asset information model (AIM); will data be successfully transferred into a computer-aided facility management (CAFM) system?

Think about how you'll wrap the data, its uses, and timing of exchanges into the contract. Ask yourself whether you should use a protocol or prescribe BIM deliverables. Ensure that you read the invitation to tender thoroughly and watch out for high levels of requirement such as LOD 500. You must consider insurance aspects. You can receive good guidance on both from the Construction Industry Council.

Also use soft landings, adopting a mind-set and a process to align design and construction with operational asset management and purpose. *Soft landings* is essentially a building completion protocol that involves gradual handover and encourages the greater involvement of designers and contractors together with building users and operators before, during, and after handover. (Refer to Chapter 9 for more about soft landings.) This alignment means that the design considers and addresses the needs of the end user throughout the design process. Designers and contractors will be involved with the building beyond its construction completion to ensure that handover becomes a smooth process, operators are trained, and optimum performance outcomes become a focus of the whole team.

Good classification systems — and in the UK, the forthcoming digital plan of work, which will help to describe data needs for a project in a simple electronic format — can aid the preceding.

Building the superstructure: Level 3

Level 3 is a sensor-rich superstructure that's constantly talking, telling you how it's performing — real-time evaluation.

Although Level 3 is still at a nebulous stage, waiting for the industry to develop all the relevant components, you can start to use some of its underpinning themes such as IFC and semantic web. You can also start to develop integrated processes and workflows within your organization with an aim of moving your data exchange lag toward being instantaneous. Try this in a collaborative workshop environment before trying it in your live project. Refer to Chapter 19 for more discussion on Level 3.

Selecting the right projects

With the right structure in place, you need to apply it to the right project. Too often people make the mistake of trying to implement BIM on a project they're already working on. Unless you've followed the BIM process from the outset, you're always going to play catch-up and you're not going to reap the rewards at the end.

You're far better to choose a project where you have a reasonable timescale, motivated supply chain, and a friendly client. Think of this project as a pilot project where you can learn, grow, and put that experience into your next project. Explaining to your client that you're adopting a new process from the outset is prudent, so as to manage any expectations. Historically, with a 2D CAD approach, a steady content supply of information was produced. However, drawings and documents produced from the BIM may come later on and in one go.

Designing to reduce change orders and variations

Regrettably, the construction industry has perversely made money from change and variation orders. BIM helps at the outset, creating a stable brief and explicit information requirements. Clients can understand the 3D model environment much more easily than a 2D drawing that only those blessed

Costing BIM-BAM-BOOM and the potential savings

The term *BIM* doesn't begin to convey the promise of cost savings over time. Think of the cost of a building over time, including its cost and operation using BIM tools. For every $1 spent in design, $20 is spent in construction and $60 in operating the buildings over its useful life of 60 years. HOK Chief Executive Officer Patrick MacLeamy devised the phrase BIM-BAM-BOOM to address the real promise of this new approach across three basic phases of a building's life. Here's what this means:

✔ **BIM (building information model):** With BIM, the 3D model investigates the available options and test performance requirements to optimize the design. BIM supports design ideas, budgeting, and program compliance.

✔ **BAM (building assembly model):** Instead of building, the contractor assembles manufactured products that are brought to site. The contractor refines the BIM to produce a BAM to allow for a reduction on construction costs. BAM allows better scheduling, facilitates subcoordination, supports cost control, and manages construction value at 20 times the cost of design. An architect and contractor team using BIM and BAM to design the building can save 30 percent of the construction cost.

✔ **BOOM (building operation optimization model):** During the building's lifetime, the owner can leverage BIM and BAM to optimize building operation. The model used in this manner is called the building operation optimization model (BOOM). BOOM helps the owner manage energy consumption and scheduled maintenance. Because BOOM is managing the value of 60 times the value of design, the cost-saving potential is enormous, saving substantially on the cost of the design and construction.

The real promise of BIM-BAM-BOOM is better design, better construction, and better operation.

with a master's in architecture or engineering can truly comprehend. The client uses the model's early doors to understand fully the proposed solutions, and makes refinements prior to procurement. The client better understanding the solutions means less risk and less modification.

Preparing for BIM Technology

Although you can't buy BIM in a box, it's a process that's enabled by technology. As well as the software itself, you need to regularly maintain and update support software, operating systems, and hardware. Having an IT strategy in place not only prevents data loss but also causes minimum disruption while the rest of the team carries out any work, meaning you can get on with the day job and still pay the bills.

As with many things in life, a successful outcome can be attributed to prior planning and thought. Before you jump right in at the deep end, take a step back and plan your approach of attack. First, know what it is that you want to achieve. In the following sections we examine the questions you should be asking to make sure you set off on the right track and are fully primed, prepped, and prepared.

Recognizing software requirements

When considering the software you need for your BIM project, make sure that your digital toolset selection is married to re-engineered and lean processes. You don't want to buy a tool and adapt your processes around its functionality. Work on optimizing your processes and procure your technology in concert with the workflow needs.

A multitude of BIM software is on the market, all with unique selling points, pluses, and weaknesses. Not only are many different software vendors with competing products available, but individual software vendors also offer more than one platform solution or suite of tools to get the job done.

Having different software that addresses different needs and requirements is particularly useful in multidisciplinary teams because no one software solution can produce a BIM project by itself. Making the right choice on any software selection should be based upon an informed decision.

Getting the right advice

Software vendors are more than willing to help you with your selection of software. Make sure that you seek advice from several software companies to compare the different software.

Chapter 22 offers plenty of resources that you can turn to when doing your research, including BIM conferences, webinars, social media, YouTube, and software user groups. Speak to your clients as well as the supply chain. Software develops at a fast pace, so be sure to review your selection periodically to make sure that it still serves your needs.

Choosing the digital equipment for your new toolkit

An asset can't be fully designed and built using software from a single software vendor, and you may consider a number of software solutions so you have the right tool for the job. Make sure that you pick the right tool for the job and not the right job for the tool.

Before selecting a platform, ask yourself the following questions:

✔ **Can I get some free stuff to support my needs?** Before you dip your hands into your pockets and part with your hard-earned cash, consider any solutions that may not cost you a dime. Chapter 21 discusses software solutions, plugins, and tools that are available for free.

✔ **Does it support interoperability?** In a nutshell, *interoperability* is the ability to collaborate, exchange, and operate on building model data, between different BIM platforms. BIM tools and platforms need to be compatible with one another at a process or workflow level, particularly when working within interdisciplinary teams that have to collaborate on the same project. Being able to communicate and exchange information with other people regardless of the software tool they're using is important. Refer to Chapter 10 where you dive a little deeper and discover more about interoperability.

✔ **Do you have an exit strategy?** What happens when a software vendor no longer continues to produce a piece of software? Do you have a game plan or an exit strategy? No guarantee exists that a software vendor will continue to produce a piece of software forever. At some point the company may move on to other products and services, leaving you unsupported. If this does occur, be prepared to make the switch to another service provider.

Software vendors keep their products updated by regularly releasing versions. These releases occur anywhere from every 12 months to three years and they often don't support backwards capability; for example, you may not be able to open a file created in the 2015 version of a software in a 2013 version. Although most programs are available on Windows, not all are available for the Mac platform.

For further information on currently certified software and the software certification scheme, see the buildingSMART website at www.buildingsmart-tech.org/certification.

Software vendors usually let you try before you buy. Normally, you follow a simple registration process and then you can download the software for a limited trial period and make your evaluation.

Considering hardware requirements

When it comes to hardware, the bigger the better is usually the case. Where possible, larger displays, more powerful computers, and fast connectivity can give you and your business the competitive edge. The same is also true of memory such as RAM and the hard-drive. However, with virtual desktops and cloud-based solutions on the rise, you may not need to increase the grunt as much as you think.

Selecting the right hardware among the technical jargon can be daunting. When considering your requirements, make sure that you cover the following:

✓ **Memory:** This refers to the area in a computer where data is stored for quick access by the computer's processor. All computers come with random access memory (RAM), which is also referred to as the main primary memory so as to distinguish it from external storage devices like disk drives. Your computer needs memory to access and run programs. The more memory you have, the more space or room you have in your computer for information and programs. As computer programs and software have increased, so too have the requirements for memory. Look for a minimum of 4GB RAM. Some software vendors recommend between 8 and 16GB of RAM for larger files.

✓ **Processing speed:** Your computer's processor speed determines in great measure the speed at which your computer runs programs or completes tasks. Think of your computer's central processing unit (CPU), more commonly referred to simply as the *processor,* as the brains of the computer, where calculations and tasks are carried out. This background processing can help performance, navigation, and the display of models. The processor speed is measured in gigahertz (GHz) where the higher the number means the faster the processor. Go for the highest affordable CPU speed rating.

✓ **Operating system:** Sometimes referred to simply as OS, an *operating system* is the software that manages your computer's hardware and software resources and provides common services for computer programs. Software vendors normally state the minimum operating systems required to run their programs and applications.

✓ **Graphics card:** Also known as a *video card,* the *graphics card* is a key component to consider.

✓ **Hard-drive:** Programs store cached data while they're operating in the background, not just when you save a file. The price of solid-state drives (SSD) is coming down, so you may want to consider a combination of a hard-drive and SSD memory.

✓ **Monitor:** Larger is usually better with monitors. If your video card supports it, you can also hook up two or more monitors, allowing you to have a program or task open on each screen. You may look like a stockbroker, but you also increase your productivity.

✓ **File storage:** Storage allows you to file information away so that you or someone on the project can retrieve it at a later date. Computers come with inbuilt storage capacity; however, network solutions such as storage area networks (SANs) and network attached storage (NAS) are becoming more commonplace. Whatever solution you choose, make sure that you also consider the issue of data security to protect your files.

✔ **Mobile versus fixed or remote:** If you travel, you may want to consider a light portable machine or a way of working remotely with your server.

Identifying training requirements

When thinking about your training requirements, make sure that you create a learning outcomes framework that sets out the needs in terms of knowledge and skills for your different levels of staff from strategic and managerial to technical persons.

Before you start, ascertain your starting position and understand the current situation in your office. A useful tip is to appoint BIM champions throughout your organization, not just at shop-floor level but right up through to senior management. BIM is as much about a cultural change as a technological one, and having BIM champions onside helps implement your strategy.

Everyone has different roles and training requirements. Just as not everyone in your organization needs to know how to create a BIM object, not everyone needs to know how to complete the BIM execution plan (BEP).

Consider your own workforce and the current skills they have. Could they become internal trainers, for example, and pass on knowledge and provide support to others? When it comes to software, instead of training everyone all at once, consider training people just in time for when they need the knowledge. You may not deploy the software for your next project for a while, at which point everyone has forgotten what they learned!

Managing BIM technology

The BIM manager may have a number of duties, such as implementing and enforcing BIM standards throughout the company as well as managing BIM technology. This may include managing software products, versions and updates, evaluating new BIM-related technology, and keeping abreast of best practice, installation, and support. When thinking about technology management, consider the following:

✔ **Undertake technology evaluations and prepare budgets.** Make sure any future technology adoption aligns with your business goals and aspirations.

✔ **Keep up to date with BIM technology.** In the fast-paced world of technology, many products and solutions only have a limited shelf life. Knowing what emerging technologies are coming to the marketplace and how they can benefit you will give you a competitive advantage. Magazines, trade shows, and the Internet are good ways to keep abreast of new and forthcoming developments. In Chapter 22 we discuss ten useful BIM resources and information sources.

✔ **Develop relationships with software vendors, resellers, and technology support staff.** Having a good support network around you and knowing who to turn to for answers may just help you out in your hour of need. Perhaps you may require additional help to get you through a technical problem or dilemma, or just want to find out little tips and tricks to assist you to get the best out of your software and processes. Don't forget to support those in your organization who may be in a satellite office or work from home.

✔ **Provide or facilitate on-going training.** This may include a mixture of internal and external training along with attending conferences, seminars, and workshops to maintain competency in your software and technology. Ask those team members who attend to bring back what they've learned and pass on information to others, perhaps by way of an internal presentation or workshop.

The terms *BIM manager, BIM coordinator* and *information officer* have crept into architecture, engineering, and construction (AEC) vocabulary. In the UK an information manager is responsible for the establishment of the common data environment (CDE) and focuses on management disciplines with no design responsibility. A BIM coordinator, on the other hand, usually contributes to information management through the establishment of model standards and execution plans.

Operating BIM in the cloud

A *cloud* is the means of storing and accessing data as well as accessing programs over an Internet connection rather than a hard-drive or local storage. Moving to the cloud means a move away from the traditional concept of purchasing dedicated hardware (which depreciates over time) to using a shared cloud infrastructure and paying as you go. You should consider cloud computing as it brings a number of benefits. These benefits can be categorized as the following:

✔ **Scalability and flexibility:** This allows IT to adapt to fluctuating and unpredictable business demands. You get the right amount of computing power you need, when you need it.

✔ **Cost savings:** In the cloud you don't have upfront hardware costs. You usually avoid the cost of maintenance and software upgrades, and you just pay for what you use. Rather than spending time maintaining your IT infrastructure, you can focus your resources, thoughts, and efforts on other business needs.

✔ **Get up and running faster:** Because you have less maintenance and no need to install software on every machine, you have time to concentrate on the job at hand. As long as you have an Internet connection, you can access the cloud from anywhere.

A number of cloud solutions are available for BIM collaboration and asset management, with the term *cloud* covering a diverse range of solutions, as the following sections explain.

As we discuss in the later section "Improving security," you do have to be mindful that you're in the hands of your cloud provider who can (in theory) access, alter, lose, or even share your data with others. Other issues exist around who owns data if it's stored on someone else's server.

Infrastructure as a Service (IaaS)

This is a form of cloud computing where virtual computing resource is provided over the Internet. The service provider offers virtual machines or storage that you can increase when required. This is one of three main categories of cloud computing services, with others including Software as a Service (SaaS) and Platform as a Service (PaaS) (see the next sections).

Software as a Service (SaaS)

Think of a bank, which protects your privacy as a customer while providing you with a service that's both reliable and secure. Web-based software, on-demand software, hosted software, pay-as-you-go software — whatever the name, what you need to know is that SaaS software and applications run on SaaS providers servers, and you access them via the Internet. You don't need to install or maintain software, and the provider manages aspects such as access, security, availability, and performance.

Platform as a Service (PaaS)

This type of cloud infrastructure is more aimed at software developers, allowing them to develop and run applications without the maintenance and complexity of infrastructure. Just like companies such as Amazon.com, eBay, and iTunes made new markets possible through a web browser, PaaS offers a fast and cost effective means of application development.

Public and private clouds

Clouds can be *private,* where you use a dedicated server for your use only, or *public,* where you share a server with others. Although a public cloud is a cheaper option, a private cloud offers greater levels of security. Increasingly, people are choosing a hybrid of the two, securing sensitive business-critical data on a private cloud and non-sensitive data on a public solution.

Considering file storage

Gone are the days where the office was littered with filing cabinets and drawing chests and lever-arch files. At the end of a project, data should be archived within the CDE for future reference. Today vast amounts of digital information are generated, and you have the opportunity to move from a physical location for storing files to a document repository system (that's well backed up).

Storage area networks (SAN) and network attached storage (NAS) are networked solutions that are becoming more popular. BIM clouds are becoming more prevalent as a way to store information over the web. As well as being a place to simply store information, many have additional functionality such as inbuilt model viewers, audit control, and instant-messaging communication systems.

For information to be trackable and easily located, files should follow a commonly accepted naming convention. Information should also provide auditable tracking of the project's history.

Improving security

Working with digital data and sharing information does bring with it some caveats. Unfortunately, criminals and other not-so-nice people may want to exploit sensitive information about an asset. So with an increasing use of, and dependence on, information and communications technologies, you need to take inherent vulnerability issues very seriously.

Make sure that information is shared and published in a security-minded fashion. Adopt a need-to-know approach to data that could potentially be exploited by baddies with a hostile or malicious intent. The loss or disclosure of sensitive information such as intellectual property may not only impact on security but could also be commercially valuable in the wrong hands. Security issues may arise around:

✔ **National security:** These issues include terrorism and organized crime.

✔ **Personal information:** Security should address any privacy issues such as the protection of personally identifiable information.

✔ **Intellectual property and commercially sensitive information:** This may include information about an asset, both physically and virtually, such as preventing data loss or information disclosure.

If your project contains sensitive information, consider implementing a policy relating to the management of access to a CDE. Inevitably, you'll create a significant amount of intellectual property to be stored in the CDE. Therefore, a policy should identify key steps to identify who can access what and how access can be granted or revoked.

Chapter 8

Setting Up a Collaborative BIM Workflow

In This Chapter

▷ Explaining the importance of collaboration

▷ Creating the employer's information requirements

▷ Considering the BIM execution plan

▷ Setting up the common data environment

---

*E*ffective collaborative working methods rely on many things such as soft skills inherent in individuals working in the project team, communication, and a desire by all to succeed. Pursuing a culture of integration and encouraging the right behaviors is vital; after all, BIM is just as much about a behavioral change program as anything else.

Effective processes and protocols are essential if the project is to be a success. Today's built assets involve a wide range of digital technologies during the whole project timeline and in producing project information. Throw the many ways and methods in which people communicate into the mix, and you can clearly see that users require clarity from an early stage if BIM is to be effective.

This chapter looks at the different elements that are required to make an efficient and effective collaborative working environment, which is right at the heart of BIM.

Getting the Lowdown on Collaboration

Collaboration is all about working together as a team to undertake tasks that achieve a shared goal. To produce better outcomes, the construction industry needs to develop better and more efficient ways of working and communicating in order to deliver better built assets. Technology and processes have evolved over time, and the construction industry has a fantastic opportunity to harness these developments in BIM. When you set up a collaborative workflow, you need to consider

- ✔ Culture and behavior
- ✔ Process
- ✔ Digital tools
- ✔ Form of contract

Figure 8-1 shows these key aspects of developing a collaborative working environment. Referred to as the Onion Diagram, it shows that you can't consider one aspect in isolation; rather, each part must have a regard to each other, with each part having an effect on the other. For example, if you consider information exchange, you use digital tools such as BIM software to produce digital data. This digital information is exchanged using defined processes that comprise managed and agreed data drops and exchanges during the project's lifecycle. To make BIM effective, all project team members need to operate in a culture and ethos of collaboration and sharing.

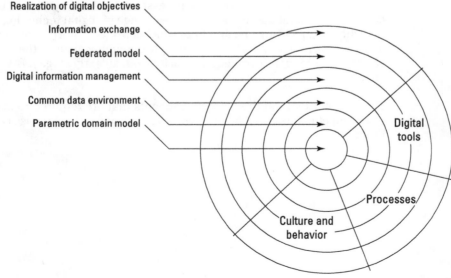

Figure 8-1:
The BIM
Onion
Diagram.

Illustration by The BIM Task Group

Modeling in the Parametric Domain

When you think about a digital model, you may initially think of its 3D geometry and what it looks like. This assumption is fair given the fact that physical architectural and construction models have been around for centuries.

However, in a BIM context the model is actually an *information model* made up of documentation, graphical information, and nongraphical information. Some of this information may live deep within the property of a *BIM object* or live in an associated document completely outside the BIM platform environment, such as a report or a survey. The following models go through a series of iterations depending on their intent.

Project information model (PIM)

During the design and construction phase, the project information model (PIM) starts out life as a design intent model, showing the designers' intentions. The project team then progressively develops the PIM across the project stages into a virtual construction model, after which the project team transfers the ownership to the construction suppliers until handover. The information manager delivers information to the employer via information exchanges. Usually, the information manager defines these information exchanges and sets them out within a scope of services document, and they correspond to important decision-making points that the client needs to action.

Include details of the arrangement and timings for the transfer of ownership of the PIM from the design team to the contractor in the conditions of engagement or contracts between the employer and supplier.

Asset information model (AIM)

The asset information model (AIM) is a single source of validated and approved information that relates to the built asset, and clients, end users, and facility managers use it for the operation and in-use phases. It may relate to a single asset, a system of assets, or the entire asset portfolio of an organization. The AIM provides a fully populated asset data set that can feed into computer-aided facility management (CAFM) systems.

The asset management team can derive information in the AIM from a number of sources, such as:

- ✔ Existing information and data transferred from existing organizational systems

- ✔ Recognizing or relabeling an existing data and information store

- ✔ New or updated information and data from surveys of the physical asset

- ✔ Information and data from the PIM

The information manager should authorize, validate, and accept information before the asset management team can rely on it as part of an AIM. The authorization and validation processes should follow any principles set out in an information management process.

Include the following in your AIM:

- ✔ Information concerning the original brief, specification, design intent, and analysis relating to the original installation of the asset and any subsequent changes

- ✔ 3D object-based models of the environment and location of the asset

- ✔ Information, or links to information, concerning the ownership of the asset and any rights or covenants associated with the asset

- ✔ Information, or links to information, concerning data obtained from the maintenance, survey, or other work carried out on the asset during its lifetime

- ✔ Information, or links to information, concerning data obtained from monitoring the operation and condition of the asset

Just like a vintage classic car, information about its past owner history and service records may be incomplete or even missing. The information manager is responsible for receiving and inputting information into the model in compliance with agreed processes and procedures. Through BIM, information can be provided to a standard data structure, which increasingly allows project teams to digitally verify that the information that is being requested has been delivered. The information manager should validate compliance with the information requirements and advise on any areas of noncompliance. An employer defines what she wants to receive as structured data, and the delivery team develops a proposal and submits a BIM in industry open standards such as IFC or COBie.

The BIM toolkit provides the tools to review the teams' submission and identifies where the client's information requirements have or have not been met.

If there are errors or omissions, the BIM toolkit will identify them and allow the user to return to their BIM tool and correct them. This saves wasted time by avoiding the client receiving incorrect submissions. To allow the BIM toolkit review process to be carried out in all types of organization, both a cloud version and a standalone, in-house version are available. After a submission is ready, it's passed on to the validation tool, which produces a simple spreadsheet report indicating the contents of a file and its compliance.

Bringing it all together

You may think of the model as a single entity, but it's actually an assembly of distinct, interlinked domain models each produced by individual contributors. These models then come together to create a complete picture of the asset. In the future you'll be working on a complete *integrated model*; however, for the time being we're talking about a *federated model*. Think of a federated model as a number of consultants' drawings, with the information subsequently transferred to a single master set.

In the following sections we discuss the central place in which the models and information are brought together, known as the common data environment (CDE) and the benefits that it brings.

Eyeing the common data environment (CDE)

The information manager carries out the process of managing these assemblies of models back and forth between the project team within one shared environment known as a *common data environment (CDE)*. A CDE is simply a place in which the project team shares information about a project, which therefore allows all information to be based upon a single source of truth. Although this information is shared and can be reused, its ownership still remains with its originator, the only person who can alter, change, or update it. This approach relies on a much more rigorous process of structuring data because others are reliant upon it. Issues such as file naming conventions and origin points become very important, but the benefits are that the project team can better use information in both construction planning and activities later in the project.

In most cases, work in progress (WIP) is within the supply chain systems. Not until information is shared or published does it go into the client's system.

After information goes through an approval stage, including a number of checks — the nature of which depends on the type of information and the point at which the project team adds it (referred to as an *approval gate*) — the information manager then makes this information available in a shared location, usually a published section of the CDE, where other members of the team can access it. The information manager may archive information that has been superseded for future reference.

A CDE could be a project extranet or a project server, and may contain useful features such as version control, search functionality, markup functionality, and the ability to control access rights through administration settings. Some CDEs can further promote the ethos of sharing and collaboration by including co-authoring functionality, survey tools, blogs, and web pages.

The benefits and advantages of a CDE are as follows:

- **Shared information reduces time and cost.** Sharing information means that another consultant or project team member doesn't have to model essential base information again.

- **Ownership of information remains with the originator.** Imagine that the structural engineer is reviewing the architectural model and she discovers that when the structural model is combined or overlaid with the architectural model, a structural column runs straight through the middle of a window. This now involves some good, old-fashioned dialogue between the architect and structural engineer as to what course of action the project team should take. Although the team shares and reuses information, only the originator of the information can change it. For example, if the structural engineer decides that the window configuration is going to change to accommodate the structural column, then only the architect can change her model.

- **The project team can generate multiple documents from different combinations of files.** Unlike traditional CAD, a section or elevation is simply a selected view from the model. Think of your digital music library that you may have on your MP3 player, phone, or computer. The information about each song is tagged with information such as artist, year, and genre, and so you can choose to listen to all country music, or perhaps Elvis is more your type of thing.

- **Version control and easy distribution provides everyone with up-to-date information.** Coordinated information means the project team is working on a single source of truth. It also means that the team isn't only working from the same information, but is also working on the latest version of that information, reducing the risk of working on an out-of-date floor plan or superseded elevation.

- **Others can use information for downstream activities.** These activities include planning, estimation, cost planning, and facilities management.

- **Information is safe and secure.** Data is stored securely, usually on a remote cloud server with monitored and controlled access.

✔ **Information can be archived.** The CDE can store previous revisions away in a digital vault, rather than the traditional physical filing and drawing cabinets. Digital information contains metadata, and so clients, end users, and facility managers can locate information quickly with a simple search function.

✔ **The postal service is relied on less, which reduces cost and saves time.** The great thing about digital uploads/downloads and electronic correspondence is that they're almost instantaneous. You can upload or download information easily at the push of a button, essentially making the postman and those expensive stamps a thing of the past.

✔ **Dispersed teams can work together regardless of where they're based.** The age of the Internet opened up a global economy, and now purchasing something from the other side of the world using your laptop and your flexible friend (credit card) is easy. The sharing of information and communication via one central place means that project teams no longer have to work in close proximity to one another. Different time zones can, of course, cause a bit of a headache, but whether you're located in London, Lyon, or Louisville no longer matters: everyone can collaborate.

Combining all the model data

Model viewing software, also known as *model integrators,* allows you to combine a number of models so that they can be viewed as a whole. Even though some software tools for collaboration and construction management have this function built in, standalone model viewers are available, many of which are free. In Chapter 11, we discuss more about open data standards and IFC, but in the meantime it's important to remember that these open standards such as IFC are important because many viewing tools use IFC information. Chapter 21 examines model viewers and checkers.

Making Life Easier with Digital Information Management

Protocols, standards, and processes are the essential foundations when setting up a collaborative BIM workflow. Other considerations to think about include the physical environment and how teams will integrate with each other; for example, co-locating teams, a cave automatic virtual environment (often just referred to as the acronym CAVE), and *big rooms* where the entire project team can come together to share best practice, knowledge, and ideas.

The world is built on standards that help drive innovation and increase productivity. On a more rudimentary level, these protocols make life easier. For example, interoperability standards such as food labeling let you clearly understand the nutritional value of the food you buy.

In the context of the digital world, standards allow you to create, use, and maintain information in an efficient way. They not only encourage best practice, but they also offer a means to achieve real improvements. Sharing construction information, drawings, specifications, and schedules in an agreed and consistent manner can bring about savings in cost and reduce waste.

In order to convey a clear message as to what digital information should be delivered, the information requirements (IR) should be documented. Then the whole project team knows what to supply and the client knows what she is getting. In the next sections, we look closer at employer's information requirements (EIR) and how the supply chain responds to this request for information via the BIM execution plan (BEP).

When the industry went from the drawing board to CAD, BS 1192 (first published in 1998) provided a guide for the structuring and exchange of CAD data. Revised in 2007 and given a new title of "Collaborative Production of Architectural Engineering and Construction Information," the standard put more emphasis on collaboration so that project team can effectively reuse data. It promoted the avoidance of wasteful activities such as waiting and searching for information, overproduction of information with no defined use, overprocessing of information simply because technology allowed it, and defects caused by poor coordination across the graphical and nongraphical data set that require rework.

Describing the digital information requirements

Information requirements need to be defined as part of the EIR, a pre-tender document forming part of the appointment and tender documents on a BIM project. The EIR sets out the information that the supplier will deliver and the standards and processes the supplier will adopt, as part of the project delivery process. The client (or the client's adviser) develops the EIR, and it's crucial to the BIM process in that client uses it to describe precisely which models she requires on a project and what the purpose of those models will be. The EIR defines which models need to be produced at each project stage — together with the required level of detail and definition. These models are key deliverables in the *data drops,* which are packets of information shared at strategic milestones along the project timeline, contributing to effective decision-making at key stages of the project and ensuring the whole project team has access to the latest information. The content of the EIR covers three areas:

- ✔ **Technical:** Details of software platforms and definitions of levels of detail
- ✔ **Management:** Details of management processes to be adopted in connection with BIM on a project
- ✔ **Commercial:** Details of BIM model deliverables, timing of data drops, and definitions of information purposes

The EIR is informed by the organizational information requirements (OIR) and asset information requirements (AIR) plus data needs to support interim decision-making.

Ideally, the EIR should advocate the use of open standards and specify the use of ISO 16739 for the exchange of model data. Doing so allows the client to select competing project teams based on her competency rather than the BIM platform being used. The compelling argument for using IFC is that newer versions always open older versions, so they're ideally placed for achieving open file formats. IFC is going to be fundamental to achieving BIM Level 3 and an environment where everyone can work from a single integrated model. This may not happen for some time, but you can start to prepare for these data transactions now.

Include the following in your plan:

- ✔ Levels of development — for example, requirements for information submissions at defined project stages.
- ✔ Training requirements.
- ✔ Planning of work and data segregation including modeling management, naming conventions, and so forth.
- ✔ Coordination and *clash detection,* which is the process of identifying conflicts and issues through 3D collaboration and coordination — also known as *interference checking.* See Chapter 17 for more information on clash detection.
- ✔ Collaboration process.
- ✔ Any specific information that should be either excluded or included from information models.
- ✔ Any constraints such as model file or attachment sizes.
- ✔ Compliance plan.

- A definition of any coordinate origin/system (three dimensions) that the employer requires the design team to use to place graphical models; for example, ordnance survey locators, geospatial, and location with respect to an agreed origin.

- Schedule of any software formats, including version numbers that the supply chain will use to deliver the project.

- Exchange of information — alignment of information exchanges, work stages, purpose, and required formats.

- Client's strategic purposes — details of the expected purposes for information provided in models.

- A schedule of any software formats, including version numbers that the supply chain will use to deliver the project.

- An initial responsibility matrix setting out any discipline responsibilities for model or information production in line with the defined project stages. The *responsibility matrix* sets out the relationships between disciplines and the production of information or models. It outlines *who* is responsible for *what*.

- A schedule of the standards and guidance documents used to define the BIM processes and protocols that the design team will use on the project.

- A schedule of any changes to the standard roles, responsibilities, authorities, and competences set out in the contract.

- Competence assessment.

The contents of the EIR may also include project-specific items such as pre-construction surveys or a requirement for the employer to receive information models describing newly generated products and assemblies.

Public sector employers may not want to (or be able to) specify software packages that their suppliers should use, but may instead specify the formats of any outputs. Private sector employers may choose to specify software packages and/or output formats.

An initial responsibility matrix forms part of the EIR, setting out any discipline responsibilities for models or information production in line with the defined project stages. The free-to-use RIBA Plan of Work Toolbox (www.ribaplanofwork.com/Toolbox.aspx) provides a downloadable spreadsheet (in Microsoft Excel format) containing customizable tables, allowing easy creation of the project roles, design responsibility matrix, and multidisciplinary schedules of services. The toolbox also contains helpful guidance and examples.

Demonstrating capability: BIM execution plans

The BEP is an important and useful document, even on a small project. Because BIM is enabling a faster design process, the need to consider software and hardware requirements, information exchanges, and how you plan to communicate with others has become even more important.

Essentially, the BEP is the supplier response to the questions and requirements set out in the EIR. Its purpose is to outline the vision and provide information and guidance to the project team to aid understanding and communication and so achieve a more collaborative process. A good plan indicates how the project team will produce and manage information and explains how the supplier will carry out and deliver the BIM project.

A number of templates and guidance documents can get you on the way to completing your BEP. The Construction Project Information Committee (CPIC) provides BEP templates for both pre- and post-contract as a free download at www.cpic.org.uk/cpix/cpix-bim-execution-plan.

The Pennsylvania State University Computer Integrated Construction (CIC) research program's BIM Planning website (http://bim.psu.edu) has developed the BIM Project Execution Planning Guide, which is available as a free download. The guide was a product of the buildingSMART alliance (bSa) Project "BIM Project Execution Planning."

To gain maximum value, make the BEP accessible to the whole project team. Put procedures in place so that any new team members are aware of the core project procedures.

The supplier prepares a BEP with the following:

- **Pre-contract:** Submitted at tender stage, pre-contract, to address issues in the EIR. It demonstrates the supplier's proposed approach, capability, capacity, and competency to meet the EIR. It provides a way for the client to assess the supplier's BIM credentials, resources, IT, and any other BIM project experience.

Information the supplier provides in the pre-contract BEP should be sufficient to enable the employer or client to review the supplier's BIM credentials so that she can make a selection or appointment. Include the following in your plan:

- **Information required by the EIR:** The BEP should respond to the parts of the EIR that specifically require a response. This is likely to include information regarding planning of work and data segregation, coordination and clash detection, health and safety management, and collaboration procedures such as details of meeting schedules and agendas, model management, including filing storage, and structure and modeling requirements.

- **Project information:** Such as project name, description, location and critical dates, key project members, project roles, company organization chart, and a contractual tree.

- **Project implementation plan (PIP):** Each organization that's bidding for a project submits this document as part of the initial BEP. Include company organization charts and details of project roles. Refer to the next section for more information about the PIP.

- **Project goals, objectives, and major milestones consistent with the project program:** Include project goals for collaboration and information management and how the supplier's proposals will meet these.

- **Project information model (PIM) delivery strategy:** Consider the delivery strategy, taking into account the deliverables, accuracy, and completeness of the design at each stage of the project from brief to in-use.

✔ **Post-contract:** Submitted after the client awards the contract to explain the supplier's methodology for delivering the project using BIM. The document should provide conformation and reassurance to the client and build upon the details and information provided in the pre-contract BEP. The post-contract BEP contains all the preceding information, plus the following:

- **Management:** Details of roles, responsibilities, and authorities, existing legacy data or survey information (if available), and details of how the supplier intends to approve information.

- **Revised PIP information exchange:** Now that the client has awarded the contract, the PIP should contain further details of the supply chain's capability to deliver the project.

- **Responsibility matrix:** It should outline the agreed responsibilities across the whole supply chain, task information delivery plan (TIDP), and master information delivery plan (MIDP). Check out the later section "Delivering the promises" for more about TIDP and MIDP.

- **Model QA procedures:** Include details of modeling standards, naming conventions, model origin points, and agreed tolerances for all disciplines and change-control procedures.

- **Delivery strategy and procurement:** Describe in detail the methods for contract management and the specifics of the procurement strategy and documentation that would need to be referenced for the actual delivery of the project.

- **Communication strategy:** Incorporate what types of meetings the project team will have and when.

- **IT solutions and technology:** Include agreed software versions, exchange formats, and details of process and data management systems.

Assessing competence: Project implementation plans

The employer uses the PIP to assess capability, competence, and experience of potential suppliers bidding for a project, along with quality documentation. It's essentially a statement that relates to the supplier's IT and human resource capability to deliver the EIR.

In this context the PIP relates to information capabilities. Don't confuse it with a generic project management plan.

The PIP should describe and give a summary of the lead party such as the main contractor as well as the supply chain's capability to deliver the project. A summary of the supply chain's capability enables the employer to make a quick assessment and comparison. The assessments should cover the following:

- ✔ **Supplier BIM assessment form:** Completed by all appropriate organizations within the supply chain. It gives an opportunity for the supply chain to highlight any previous BIM experience and what benefits they realized, as well as the opportunity to demonstrate that the supply chain understands the analysis and methods that are proposed for the project.

- ✔ **Supplier information technology assessment form:** Completed by all appropriate organizations within the supply chain, usually with the assistance of the organizations' IT departments. It demonstrates the capability and IT resource of the supplier for information exchange in a collaborative environment.

- ✔ **Supplier resource assessment form:** The supplier uses this to assess the organization's current resource capability and capacity to demonstrate its capability to deliver the project.

The principal supplier is responsible for obtaining enough information about the supply chain to assure that it has the appropriate capability to meet the requirements set out in the contract and the EIR. Resolve any potential problems or issues around interoperability and information exchange as early as possible, ideally before the design commences. If the IT assessment highlights an incapability, in that files can't open, read, or analyze models from the different teams, then drawing production will be difficult to achieve.

A simple BIM capability questionnaire is a good way to help the project team identify any training, coaching, or support requirements that the supply chain may require. The CPIC provides a BIM assessment form as a free download (www.cpic.org.uk/cpix/cpix-bim-assessment-file) that includes a BIM Capability Questionnaire template with 29 questions. The questions were created as part of working documentation provided by the Skanska UK BIM team.

Delivering the promises

After the client awards a contract, the project lead usually calls an initial meeting to collaborate and develop the MIDP made up from the team members' TIDP. The following sections take a closer look at these two forms.

Master information delivery plan (MIDP)

The main contractor uses the MIDP to manage the delivery of information during a project. It lists the following information deliverables for the project:

- Models
- Drawings
- Specifications
- Equipment schedules
- Room data sheets

Task information delivery plan (TIDP)

The TIDP is a federated list of information deliverables that are broken down by each task, including format, date, and responsibilities, and it forms part of the BEP. The main contractor should use the information within the TIDP when working out the required sequence of model preparation for any work packages used in the project.

The contractor uses the TIDP to

- Demonstrate what team member is responsible for each task or deliverable.

✔ Indicate that milestones within each TIDP are aligned to design and construction programs to produce the MIDP.

✔ Show that each task team manager is responsible for compiling her own TIDP with its milestone.

Singing Off the Same Hymn Sheet: Information Exchange

Having a protocol in place sets the foundations for everyone to work collaboratively and exchange information. Protocols can also put in place robust quality assurance (QA) procedures.

In response to the UK government's BIM strategy, the Construction Industry Council has published the CIC BIM Protocol (together with other supporting documentation). This free-to-download guide (`http://cic.org.uk/publications`) is concise at only eight clauses long, but highlights the minimum legal and commercial requirements that apply when using BIM on a project, clarifying rights, responsibilities, and liabilities of each party involved. It also features a prototype production and delivery table to clarify which models the client requires at different stages of a project, who's responsible for the information at each data drop or work stage, and the required level of detail.

Part III
Understanding BIM Requirements and Developing BIM Processes

Level of Detail (US)	Level of Detail (UK)	Geometric Detail	Level of Information
100	1	Approximate dimensions and concept graphics for visualizations.	The project requires a boiler. Initial options described.
200	2	Generic boiler system, space arrangement, clearance zones, and input/output.	System description and initial requirements specified for performance of boiler.
300	3	Development of detail, dimensional constraints for technical design, materials, and finishes.	System selection and detailed performance requirements with reference to standards.
400	4	Technical design proposals, detailed spatial coordination of related systems.	Technical system specification, including components to permit product selection.
500	5	As-built installation and coordination, future use as maintenance information.	Detailed specification of manufacturer's product, including testing, operation, and maintenance.

Prepare yourself for implementation by finding out the BIM fundamentals you need to put in place at www.dummies.com/extras/bim.

In this part . . .

✔ Review where BIM is becoming a requirement around the world, and understand the key BIM mandates, standards, and protocols so you can adopt and implement a robust BIM strategy within your organization.

✔ Develop high-quality BIM objects with robust data properties and accurate geometry that support the whole project lifecycle.

✔ Recognize that you need to use shareable formats and free information to support BIM's openness and collaboration, and be able to work on a BIM project regardless of different software being used.

✔ Revolutionize your business by encouraging BIM champions and providing the right levels of training to transform your workplace and give the team the support and encouragement it needs to succeed.

✔ Count the cost of BIM implementation and see how you can make a return on that investment, increase efficiency, and improve timescales.

✔ Deal with the legal aspects of BIM methods and control your liability, risk, and intellectual property to ensure that you and your business thrive.

Chapter 9

Mandating BIM in the UK and across the Globe

*B*IM is being adopted around the world, governments are providing financial stimuli and creating mandates for BIM deliverables, and project delivery methods are changing. It's not a matter of if, but *when* your firm will implement BIM. You can look at the UK as a great example of how turning BIM into a national policy accelerates adoption. In its first three years, the UK Construction Strategy saved the government £1.4 billion ($2.1 billion).

The UK government's decision to mandate BIM has led to more engagement from the entire supply chain, including clients and owner-operators, than in the United States. The United States is arguably behind in terms of BIM maturity and a centralized approach, but huge potential exists for standardization. In this chapter, we take a look at the UK government's Construction Strategy and where the BIM mandates came from. We also take a high-level look at the US picture and consider other BIM mandates across the rest of the globe.

Interpreting the Requirements

At the end of 2014 a major development occurred. The European Commission endorsed the formation of the European Union BIM Group and gave financial and secretariat support to the group. EU BIM will be funded to mid-2016 and likely deliverables include reports for the introduction of digital technology and BIM to European public works, as well as communications and events to engage the public client community and support the push for the private sector.

In March 2014, the EU published the European Union Public Procurement Directive (EUPPD) in a bid to modernize the existing EU public procurement rules by simplifying the procedures and making them more flexible. The directive makes a specific reference to the use of BIM in public works in Article 22(4):

> *For public works contracts and design contests, Member States may require the use of specific electronic tools, such as of building information electronic modelling tools or similar . . .*

In simple terms, EU member states have until 2016 to implement BIM policy into their national legislation. Making BIM policy a requirement not only encourages individual nations to develop a digital construction industry, but also makes collaboration between the EU member states more likely.

The global construction industry clearly can take away much from looking at different BIM mandates around the world and potentially reap the rewards and benefits that a standardized approach to BIM brings. The following sections look a little deeper into the UK government BIM mandate, its origins from the UK government Construction Strategy, and the value proposition the UK government believes BIM can bring to the economy and construction industry.

Knowing where to start with the UK BIM mandate

The decision to adopt BIM is an individual one for every business. Making the decision is more complex than if you were switching from a drawing board to CAD, but the adoption is definitely worth the effort to reap the rewards. The UK government realized that it wasn't using the wealth of construction and real estate data available in an efficient way. By analyzing data more effectively, it was possible to understand more about the UK building portfolio and how the industry could evolve.

Think about the retail sector for a second and the moment you hand over your hard-earned cash to a merchant in exchange for goods or services. Once, the point-of-sale system typically included a cash register, a receipt printer, a barcode scanner, and a monitor and customer display. The new electronic point-of-sale systems (EPOS) cover the basic functions of a till such as calculations and issuing receipts, but make more efficient use of data.

For example, they can work out which product lines aren't doing so well, link to stock levels, and store information about you, the customer. Perhaps you've wondered why you received a discount voucher for a 20-pack of diapers and shaving cream? From the contents of your shopping cart and along with that loyalty card you signed up for, supermarkets know more about you than you think.

In the UK, the government has been the main driving force in creating a whole sector pull for BIM adoption. This is changing rapidly, though, as the construction industry reaches the tipping point where most organizations are doing BIM anyway to unlock more efficient ways of working. A mandate isn't the same as forcing companies to do something they resist; it's a catalyst designed to encourage digital innovation in a certain part of the industry — in the UK's case, public sector contracts — but involvement remains a choice for an individual company because it could focus on other areas of the industry, private house-building for example. Eventually, as more of the supply chain uses digital data exchange, the piece of the pie available to BIM-resistant companies shrinks until it's impossible to maintain as a business proposition.

Construction strategy

In 2011, the UK government published its Construction Strategy aimed at reducing the cost of public-sector assets by up to 20 percent by 2016. The strategy calls for "a profound change in the relationship between public authorities and the construction industry to ensure the Government consistently gets a good deal in the capital/delivery phase of projects."

The Construction Strategy put the spotlight on an efficient construction industry as being vital to the UK economy, because it represents approximately 7 percent of gross domestic product (GDP), or £110 billion ($170 billion) per year.

The BIM mandate's focus on public sector is because it contributes 40 percent of that figure and government is the construction industry's biggest client. Overall, the UK needs to reduce the cost of public construction projects so the government can then reinvest savings into other centrally procured projects.

The following excerpt from the 2011 Construction Strategy highlights the UK government's concern over the failure of the construction industry to take advantage of innovative digital technologies. It also demonstrates the government's understanding of the main reasons for that failure and its analysis of the resultant loss to the industry and its clients.

> ✔ *2.29 Construction has generally lagged behind other industries in the adoption of the full potential offered by digital technology.*
>
> ✔ *2.30 A lack of compatible systems, standards and protocols . . . [has] inhibited widespread adoption of a technology which has the capacity to ensure that:*
>
>   • *All team members are working from the same data.*
>
>   • *The implications of alternative design proposals can be evaluated with comparative ease.*
>
>   • *Projects are modelled in three dimensions, eliminating coordination errors and subsequent expensive change.*
>
>   • *Design data can be fed directly to machine tools, creating a link between design and manufacture and eliminating unnecessary intermediaries.*
>
>   • *There is a proper basis for asset management subsequent to construction.*

Among other objectives, such as elimination of waste, new methods of procurement, value for money, and benchmarking, BIM was high on the agenda. The report announced the government's intention to require collaborative 3D BIM on all its projects by 2016. It was clear that all information and documentation must be in electronic formats, and that it included renovations to existing estates as well as new-build projects.

The requirement in the UK specified BIM and electronic data exchange. However, you shouldn't treat BIM in isolation, but as part of a bigger change agenda such as Lean working practices, a system for eliminating waste and adding value in production processes like construction that originated at Toyota factories in Japan.

Wider benefits of the government's strategy

The objective of the government's strategy is to fast-track BIM adoption throughout the UK construction industry, but that's because it has direct benefits on construction projects, most obviously significant cost savings. Understand that BIM can't produce results such as efficiency and cost savings on its own; rather BIM is part of a much bigger picture that includes the following:

✔ **Intelligent clients:** Know what they want, what it should cost, and how best to go to market.

✔ **Transparent pipelines of work:** The UK government has been clear about the funding that it has allocated for work that is in either in the process of being developed, provided, or completed.

✔ **Early contractor engagement:** The UK government has been experimenting with new forms of procurement that involve early contractor involvement to support BIM. However, of equal importance is early engagement of specialist or subcontractors, plus manufacturers who in most cases actually create the data for the construction and asset models.

✔ **Benchmarking/use of data:** To measure efficiency and value of your data, you need a benchmark, which acts as your standard starting position that suits your organization and from which measurements can be taken. You especially need to know whether a project represents good value for money and clients need to have some idea of what projects should cost.

✔ **Lean delivery:** The two concepts of Lean construction and BIM aren't reliant on each other. However, only when project teams realize the benefits of both in combination will the construction industry revolutionize the way it works. Undoubtedly, an understanding of Lean insights and the adoption of appropriate Lean principles enhance an organization's BIM implementation processes, and, equally, an understanding of the power of BIM helps individuals and organizations to implement their Lean strategies.

✔ **Designing for manufacture and assembly (DfMA) and build off-site (BOS) standardizations:** The creation of standard digital libraries is helping to drive down capital expenditure costs and speed up delivery cycles through a build off-site agenda. Designing with accurate assembly in mind and taking more construction processes off-site directly reduces waste.

✔ **Government soft landings:** A *soft landing* is about having the approach and plan to always keep operational needs, including the client's original purpose, in mind all the way through design and construction. This means that all user's needs will be picked up in the design and resolved in the final built asset. According to the UK government, soft landings is the "golden thread" joining the capital and operational stages. Refer to the next section for more on soft landings.

A building's for life, not just for construction

One of the central themes of the government's Construction Strategy was ensuring that clients get the assets they want, meeting their needs, and performing to the level that was planned in the brief. In support of this strategy, soft landings was mandated and powered by BIM, whereby the project team considers the end user's needs from the outset and right the way through the design lifecycle, engaging the facilities management (FM) team from the start. Rather than handing the keys to the owner at the end of the job and making a quick getaway, designers and contractors are involved for an extended period of time post-construction and handover to iron out any wrinkles and make sure that the building is running through post-occupancy evaluation and refinement.

Pretend for a moment that you're in the market for a new car. You see your perfect model. With leather seats and a top-of-the range entertainment system, not only are you going to make Bob next door jealous, but your golf clubs fit perfectly in the trunk. The car comes with a manual showing you everything you need to know about your new purchase, from how to adjust seat positions to what pressure the tires should be inflated to. The garage has access to a full digital record of the technical data on each material, component, assembly, and system in the car. They can understand the intended operation of the engine management systems and identify and replace any part simply by referring to the data. What's more, they have access to guidance on how to carry out any maintenance procedures.

When you tell the manufacturer about any complaints, it accepts your feedback and either takes corrective action by recalling the vehicle for modification or making sure that future models overcome the problem so that future sales figures don't dip. If that's what you get when you buy a car, why wouldn't clients want the same when buying a building, a rail system, a motorway, or a flood alleviation scheme? Acquiring a new asset should be just as positive an experience; unfortunately, there is often a significant gap between the client's expectations and those of the design and construction teams.

The government believes that better outcomes in design and construction can be achieved by powering soft landings with BIM. You find the true value of built assets during their operational lifecycle, and soft landings helps you achieve that.

Soft landings rely on the following:

- ✔ **Early engagement:** Get both the client and end users involved early and include a soft landings champion on the project team during the design/construction process to ensure that operational considerations are included at every stage. As project time passes, the ability to make valuable change declines and the cost of change rises. The cost of operating and maintaining a building for more than 20 years can be up to three times the construction cost, so early engagement really becomes a no-brainer.

- ✔ **Aftercare commitment:** A guarantee of post-construction assistance from the project team is essential. Aftercare is about fine-tuning, making sure building systems run smoothly and ensuring the end users have a good understanding of their working environment during a measured handover period.

✔ **Feedback:** You can use post-occupancy evaluation (POE) for clients and operators to provide feedback to the project team so that you can capture lessons for the future. POE should be part of standard procedure on all your projects.

BIM goes hand-in-hand with soft landings because the collaborative way of working enables early engagement with the client. The use of a 3D model is beneficial not only for visualization and communication purposes but also for lifecycle model testing at the pre-construction stage. BIM also provides a dataset from an asset information model (AIM) that a facilities manager can use for a computer-aided facilities management (CAFM) system.

A great place to start is the BIM Task Group website at www.bimtaskgroup. org. Its aim is to support the main objective of the government's Construction Strategy — achieving Level 2 BIM by 2016. The website provides up-to-date news on the group's program, links to key resources, and access to lessons-learned documentation from exemplar projects.

Construction Strategy 2025

In 2013, the government published the Construction 2025 — Industry Strategy for Construction, which sets out a long-term vision and action plan by government and industry to work together to promote the success of the UK construction sector, including the benefits of BIM. The main objects are to

✔ **Reduce costs.** The construction industry must lower costs by 33 percent in both the cost of construction and the whole-life cost of built assets.

✔ **Build faster.** The construction industry must deliver faster, finding a 50 percent reduction in the overall time it takes from client request to project completion, for both new build and renovation of assets. This will be enabled by a move toward automation and designing for manufacture and assembly (DfMA).

✔ **Be kinder to the environment.** The construction industry must lower greenhouse gas emissions by 50 percent.

✔ **Improve exports.** The construction industry must see a 50 percent reduction in the trade gap between total exports and total imports for construction materials and products. You need to think more about how you build commercial opportunities around your datasets.

Through the implementation of BIM, the construction industry can meet these challenges and deliver more sustainable buildings, more quickly, and more efficiently.

Ready steady: You push, I'll pull

The BIM Industry Working Group report "BIM Management for Value, Cost, and Carbon Improvement" looks at the construction and post-occupancy benefits of BIM for use in the UK buildings and infrastructure markets. The report recommends a *push-pull* strategy, which just means pushing the supply chain to achieve a minimum level of BIM use, and pulling from the client-side by requesting information and actually making best use of the data collected. These sections examine the push-pull aspects of this strategy in a bit more detail. (To read the report in full, visit www.bimtaskgroup.org/wp-content/uploads/2012/03/BIS-BIM-strategy-Report.pdf.)

Supply-chain push

The strategy supports the push element that aims to help the supply chain make BIM use simpler, including providing guidance about BIM processes and training resources about the deliverables required. You have seen software vendors who advertise that they offer the best BIM solution and members of the supply chain who indicate that they have the BIM method to rule them all.

The strategy set out to produce a package of standards and resources that support straightforward delivery of BIM Levels 2 and 3. BIM maturity can be measured along the Bew-Richards maturity model, which the UK industry commonly refers to as the BIM wedge. Refer to the "Exploring the BIM wedge" section for more information. In helping with the widespread adoption, the government won't stop innovation or the freedom of the members of the supply chain to make their own decisions about the systems and software packages that they use.

Client-side pull

The flip side to the push is the pull. The government as a client has a strategy to be clear about what data it will buy and use and to standardize this wherever possible. The government as a sponsor of the construction industry has an interest in maximizing the benefits of BIM, and it needs to be specific about its own requirements. For example, it has a responsibility to set out the data deliverables necessary at key stages in the project and lifecycle of a built asset, which means digital submittals and handover information can be checked consistently on every project.

Including BIM in national policy has accelerated adoption in the UK. Contact government representatives to start encouraging the same legislative change in your local region.

Exploring the BIM wedge

The UK government has used a diagram to clearly indicate a target of Level 2 BIM in the UK maturity model, affectionately known as the *BIM wedge* due to its shape. The UK maturity model, developed by Mark Bew and Mervyn Richards, has become an integral part of any corporate presentation on BIM. It allows

- ✔ The construction supply chain to clearly identify what it is to deliver
- ✔ The client to precisely understand what the supply chain is offering

Figure 9-1 depicts four levels, as follows:

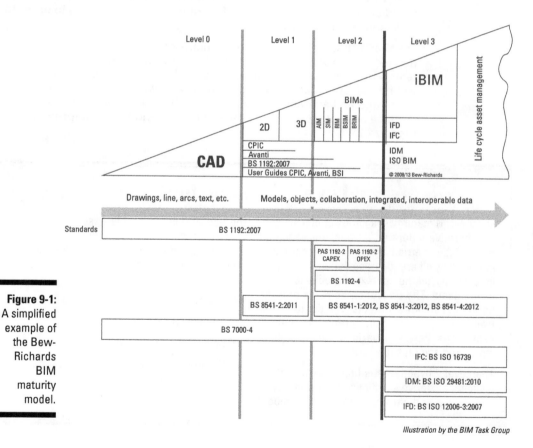

Figure 9-1: A simplified example of the Bew-Richards BIM maturity model.

Illustration by the BIM Task Group

- ✔ **Level 0:** Think of outputting 2D CAD information, consisting of lines, circles, and text, as paper pages and electronic printouts. At this level no collaboration is taking place.

- ✔ **Level 1:** This is usually demonstrated by a mixture of 3D and 2D CAD production. Some file-based collaboration may be occurring, with you sharing data via document management, but generally no collaboration is taking place between different disciplines.

- ✔ **Level 2:** This is defined as file-based collaboration and library management and is a series of domain-specific models (architectural, structural, services, and so on) within a single environment where the project team can share structured data based on Construction Operations Building information exchange (COBie). It's generally based on historical data, rather than real-time information, so it relies on shared content, which could be work in progress, being federated with other inputs in order to make informed decisions.

- ✔ **Level 3:** Although the exact requirements of Level 3 are yet to be defined, this is about fully integrated BIM (or iBIM) where the project team works on one central model somewhere in the cloud. Level 3 will rely on open-data standards and move toward *real-time data* (an integrated and automated environment where the project team will be working on a single model, at the same time) and look at improving organizational performance.

Level 2 in a nutshell

Level 2 BIM maturity is a series of domain and collaborative federated models, consisting of both 3D geometrical and nongraphical data, prepared by different parties during the project lifecycle within the context of a common data environment. The project participants provide defined, validated outputs via digital data transactions using proprietary information exchanges between various systems in a structured and reusable form.

The Level 2 hypothesis is that government as a client can derive significant improvements in cost, value, and carbon performance through the use of open, sharable asset information. BIM ensures that clients are buying their data only once, and that they're using that data to make earlier and better-informed decisions. The key Level 2 requirements are

- ✔ The exchange standard of COBie

- ✔ 2D reviewable PDF

- ✔ Copies of native file formats

- ✔ Standardized approach to information management

When the industry talks about BIM, it's talking about computer-readable data that people can use to make smart decisions. With artificial intelligence, will the computer make decisions for you in the future? As you move toward BIM Level 4 and beyond, the industry is thinking about behavioral datasets. For example, what effect would more natural lighting have on patient recovery time or school grades? The project team will start to simulate these kinds of outcomes, beyond traditional facilities management. If you're interested in this, take a look at Chapter 19.

Building a framework

The BIM Task Group, Construction Industry Council (CIC), and UK British Standards Institute (BSI) have produced a suite of documents that give industry the tools, processes, and procedures to work at Level 2 BIM.

In the UK, the BSI is the primary publisher of standards-based documentation. Through its technical committee, B/555, the construction design, modeling, and data exchange, the BSI has been developing standards to support BIM. The BSI (B/555) roadmap, created in conjunction with the BIM Task Group, gives an overview of the committee's activities that provide support in delivering clear guidance to the UK industry.

UK Level 1 standards

The UK Level 1 standards support Level 1 of the UK BIM maturity model. The standards are centered around the collaborative production of 3D and 2D CAD information and guidance on managing the design process. Table 9-1 lists the Level 1 standards.

UK Level 2: Processes, tools, and guides

The UK Level 2 standards are standards, procedures, and supplementary documents that help you achieve BIM Level 2 compliance, which is known as collaborative BIM, where the project team use their own 3D models, but don't necessarily work on a single, central model.

Table 9-1	Level 1 Standards
Title	*Description*
BS 1192:2007	Collaborative production of architectural engineering and construction information.
BS 7000-4:2013	Design management systems, part 4; guide to managing design in construction.
BS 8541:2:2011	Library objects for architecture, engineering, and construction — recommended 2D symbols of building elements for use in BIM.

The documents support collaboration by describing how data is exchanged between the members of the project team. Level 2 collaboration often uses a common file format to share information, which means any company can merge project content with their own to create a federated model and run checking and validation tools on the BIM. Table 9-2 lists the Level 2 documents.

Table 9-2	Level 2 Standards
Title	*Description*
PAS 1192-2:2013	Specification for information management for the capital/delivery phase of a construction project using BIM.
PAS 1192-3	Specification for information management for the operation phase of assets using BIM.
BS 1192-4	Collaborative production of information part 4; fulfilling employer's information exchange requirements using COBie — code of practice.
PAS 1192-5	Specification for security-minded BIM, digital built environments, and smart asset management.
CIC BIM protocol	Standard protocol for use in projects using BIM.
Soft landings and compendium 8356:2015	Graduated handover of a built asset from the design and construction team to allow structured familiarization of systems and components, and fine-tuning of controls and other building management systems.
Classification	Systematic arrangement of headings and sub-headings for aspects of construction work including the nature of assets, construction elements, systems, and products.
Digital plan of work (dPow)	Generic schedule of phases, roles, responsibilities, assets, and attributes, made available in a computable form.

Level 3 standards

The concept of Level 3 is centered around the idea of a shared, integrated BIM, also referred to as *iBIM*, and is yet to be fully defined in terms of detailed standards, but more clarity will emerge during the development of Digital Built Britain, the group responsible for helping the UK reach BIM Level 3. Refer to the next section for more information.

The standards and processes around Level 3 BIM maturity will develop over the next few years from a technical, commercial, and indeed behavioral perspective. There'll also be new forms of procurement: integrated and performance based. Key to this maturity will be service performance data, such as energy use on a wide scale, enabling smarter decisions to be made by clients and operators at all levels.

Table 9-3 highlights the open standards that will steer the construction industry toward BIM Level 3.

Table 9-3	Level 3 Standards
Title	*Description*
ISO 12006-3:2007	Building construction — organization of information about construction works — part 3, the framework for object-oriented information (IFD).
ISO 16739:2013	Industry Foundation Classes (IFC) for data sharing in the construction and facility management industries.
ISO 29481-1:2010	BIM information delivery manual (IDM) — part 1; methodology and format.
ISO 29481-2:2012	Building Information Modeling — information delivery manual (IDM) — part 2: interaction framework.

Securing the future — Digital Built Britain

The Level 2 BIM program is the main enabling strategy for transforming UK construction into a digital industry, but don't get out of the elevator just yet. The industrial strategy for construction announced that the UK is going to be "an industry that is efficient and technologically advanced." Digital Built Britain (DBB) is the brand that will help the UK deliver BIM Level 3 capability to domestic and international markets and link BIM to smart cities and smart grids. Check out www.digital-built-britain.com.

UK BIM survey results suggest the industry is now reaching a stage where BIM is becoming the norm, although smaller practices are lagging behind their larger competitors. The UK construction industry sees adoption of BIM as bringing competitive advantage and that the current UK government's BIM targets are achievable.

For example, HS2 Ltd, the company responsible for developing and promoting the UK's new high-speed rail network, undertook a study to test whether the supply chain is ready to work to BIM Level 2. The study found that 94 percent of the supply chain is already using BIM and that 60 percent has a BIM strategy with defined goals.

Leading BIM in the United States

You can validly think of the United States as the founding fathers of BIM theory, being the home of parametric CAD innovators like Chuck Eastman at the Georgia Institute of Technology and Bill East of the US Army Corps of Engineers (US ACE) who devised the COBie data exchange structure. Even the successful BIM policy in the UK held up as a global exemplar was born out of development discussions with the US General Services Administration (GSA). Since 2003, the US federal government has expressed its BIM ambition, and the American Society of Civil Engineers proposed in 2009 that all road projects be built in BIM. However, very few national BIM strategies exist except those that are focused in the military.

The GSA has played a vital role in promoting BIM within the construction industry, being the first organization to lead the US government into BIM. Other government agencies such as the Naval Facilities Engineering Command (NAVFAC), US Coast Guard (USCG), US Air Force, and US Army are also using BIM. However, although BIM adoption is high in the United States, no central standards or nationwide organizing principles exist. Unlike the UK, the US government isn't as responsive and the construction industry is spread out across states with unique requirements.

In the following sections you take a closer look at some of the good documentation, standards, and guides that the likes of the GSA, US ACE, and the National Institute of Building Science (NIBS) have produced that you can also use to your advantage.

Reviewing the GSA BIM guides

Since 2006, the GSA has mandated that those working on a new building designed through its Public Buildings Service (PBS) shall use BIM at the design stage. Since 2007, those receiving design funding must use BIM for spatial programming as a minimum requirement on all major projects when submitting to the Office of Design and Construction (ODC) for final concept approvals by the Public Buildings Service (PBS) commissioner and the chief architect. You can find out more about the GSA National 3D-4D-BIM Program at www.gsa.gov/portal/content/105075.

The GSA has developed the following suite of BIM guides and documentation:

- ✔ **GSA BIM Guide 01-Overview:** This document is a good starting point and acts as an introductory text, providing the foundations and a common starting point. It introduces the GSA's National 3D-4D BIM program and provides background information on the concept, definition, and expectations underlying 3D and 4D BIM technologies and models.

- ✔ **GSA BIM Guide 02-Spatial Program Validation:** This guide describes the tools, processes, definitions, and requirements to use BIM effectively for GSA's spatial BIM minimum requirements. The GSA aims to leverage the use of BIM and interoperability to automate the checking of model integrity and design performance.

- ✔ **GSA BIM Guide 03-3D Laser Scanning:** With 3D laser scanning becoming a valuable way in which to obtain spatial data, the GSA is currently encouraging, documenting, and evaluating laser scanning technology on a project-by-project basis. This guide is currently in draft and contains good advice when contracting for and ensuring quality in 3D imaging contracts.

- ✔ **GSA BIM Guide 04-4D Phasing:** The focus of this document is on describing the tools and processes required to explore how phasing and time-related information will affect project development.

- ✔ **GSA BIM Guide 05-Energy Performance:** This guide describes the use of space-based BIM modeling techniques and strategies to strengthen the reliability, consistency, and usability of predicted energy use and cost estimates during the design and operation phases.

- ✔ **GSA BIM Guide 08-Facilty Management:** GSA understands that there is great benefit in using and maintaining data throughout the lifecycle of facilities. This guide describes how this data generates efficiencies such as reducing the cost of renovations and optimizing building systems to reduce carbon and energy usage.

Viewing the NBIMS-US

The National BIM Standard-US (NBIMS) is a consensus document and an on-going initiative of the buildingSMART alliance, a council of the National Institute of Building Science, which focuses on providing standards that facilitate efficient lifecycle management of the built environment with support from digital technology.

It references existing standards, documents information exchanges, and delivers best business practice for the entire built environment. The document is built upon open BIM standards such as Industry Foundation Classes (IFC) and the scope covers BIM execution plans (BEP) and data exchange.

The first version of the standard was initially released in 2007 with an updated version following in 2012. Keep an eye out for Version 3 of the standard that will be available at www.nationalbimstandard.org. The NBIMS project committee is always looking for a cross-section of the construction industry that wants to get involved. You can refer to its website for further details.

Following the Veterans' Affairs (VA) BIM guide

The US Department of Veterans Affairs (VA) Office of Construction & Facilities Management (CFM) provides a number of services to the VA to deliver facilities in support of US veterans. The VA found that digitizing its patient medical records generated process efficiencies in administration and management. The CFM now uses BIM to support similar process improvements in the built environment.

The VA BIM guide aims to support the adoption of BIM on relevant projects including help and guidance on the size and type of projects that BIM is most effective for. The guidance is intended to apply to a range of skill sets and discipline sectors.

You can access the VA BIM guide either as a PDF download or by using the interactive website at www.cfm.va.gov/til/bim/BIMGuide. The guide covers aspects such as design, construction and implementation, roles and responsibilities, modeling requirements, and security.

Remembering COBie and the US ACE

Construction Operation Building information exchange (COBie) started in the United States as part of the US Army Corps of Engineers. The Corps now produces additional CAD and BIM resources through the US ACE CAD BIM Technology Center at https://cadbim.usace.army.mil. The center's guidance covers the application of BIM to facility management, planning, and civil engineering. The resources are organized around the US ACE BIM Roadmap and include information on BIM contract requirements. The site also includes an AEC CAD Standard and templates for Autodesk Revit and Bentley Microstation.

In the true spirit of collaboration, the center also includes a workspace for what is called the Tri-Service (made up of US Army and US ACE, Naval Facilities Engineering Command (NAVFAC), and the US Air Force), designed for military construction and civil works project teams including firms under contract as part of the wider supply chain.

Growing BIM Standards Internationally

More and more countries around the world are embracing BIM. Although some countries are creating a top-down approach by mandating BIM at a government level, other countries are using a bottom-up approach. Figure 9-2 shows the location of various BIM mandates around the world. The countries with wave shading have mandated BIM, and those with solid, black shading are on the road to BIM adoption and national policy.

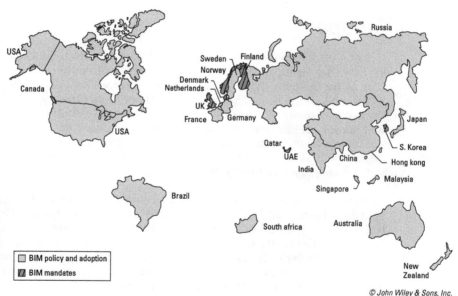

Figure 9-2: Map of global BIM mandates.

☐ BIM policy and adoption

▨ BIM mandates

© John Wiley & Sons, Inc.

The following sections take a high-level look at global BIM policy, introducing you to some great ideas, comparing international activity, and providing you with starting points to investigate BIM adoption in these territories.

Reviewing global BIM mandates

The UK is the only nation with a public-sector mandate. However, the private sector is keen to work with BIM-enabled teams, and private clients and contractors are beginning to mandate BIM on particular projects, in countries where government mandates are in place and those that haven't yet adopted BIM. Although the UK's global lead may not last, work has begun on the data-sharing guide PAS 1192:2 as an ISO international standard, making its impact felt globally.

The 60-page McGraw Hill "Smart Market Business Value of BIM for Construction" report gives a good insight and overview of the adoption and use of BIM worldwide. The report focuses on the largest global construction markets, including Australia, New Zealand, Brazil, Canada, France, Germany, Japan, South Korea, the United States, and the UK. The report gives further insight with market and profile of contractors using BIM, the benefits and return on investment (ROI) derived from their BIM investments, and the critical activities and practices for which they're using BIM.

Europe

Demand for BIM is growing in continental Europe, with the formation of the EU BIM Task Group and implementation of BIM into public works legislation, as we describe in the earlier section "Interpreting the Requirements." If you think of the United States as the fathers of BIM theory, then Europe is the home of many of the software vendors that have evolved CAD tools into BIM platforms, like Nemetschek (Germany) and Dassault (France). As well as the UK government, many other client bodies have established BIM requirements, such as Germany's Deutsche Bahn Transport and the Netherlands' water management ministry, Rijkswaterstaat.

Fourteen countries have indicated particular enthusiasm for Europe-wide specification of BIM: Austria, Denmark, Estonia, Finland, France, Germany, Iceland, Italy, the Netherlands, Norway, Portugal, Slovakia, Sweden, and the UK. France, Germany, and Norway, in particular, are starting government-led initiatives similar to that in the UK. Others, like Sweden and the Netherlands, have put the focus on particular agencies to develop requirements with industry. Central and Eastern Europe is building an impressive portfolio of technology companies and a digitally advanced workforce to facilitate BIM from practice.

The following list gives a brief overview of the BIM activities taking place across Europe:

✔ **Denmark:** A national mandate from the Ministry of Climate, Energy, and Building exists in the form that Danish state clients such as the Palaces and Properties Agency now require their supply chain to use BIM for projects over DKK5 million.

✔ **Finland:** Established in 2012, the Common BIM Requirements BIM to be used for all state property and national public projects. Finland is often held up as an exemplar of BIM adoption and early innovation, because of its relatively small industry and high levels of technology and open data exchange.

✔ **France:** A task group has been established to develop a BIM mandate first put forward by the Ministry of Dwellings (Ministère du Logement). With a budget of €20 million ($22 million), the group is taking plans announced in June 2014 to build 500,000 BIM-developed houses by 2017. To take the housing ambition forward, Le Plan Transition Numérique dans le Bâtiment (the Digital Building Transition Plan) task group has been formed. Also, the French National research Project MINnD has been set up, and the consortium joins together contractors, engineers, software vendors, academia, and professional institutes. The "Interoperable Information Model for Sustainable Infrastructures" project has commenced to develop and explore open BIM standards for infrastructure projects. See more at `www.minnd.fr/en`.

✔ **Germany:** The Federal Ministry of Transport and Infrastructure has led the creation of an industry "Digital Building Platform," in coordination with industry organizations, with a key focus on standardization, digital data exchange, and new BIM-ready legal contracts. The strategy is part of a reform commission for construction that looks to understand why a number of large German public sector projects had significant time and cost problems.

✔ **The Netherlands:** In combination with many other European countries, especially for infrastructure, BIM has been developing in the Netherlands for many years. In 2012, a Building Information Council was set up as part of the Rijkswaterstaat highways and waterways BIM program. There are many aspects to the project, including a standard format for data exchange and cross-discipline translation.

✔ **Norway:** Statsbygg, which is Norway's public and civil client, developed a phased approach to mandating BIM, putting frameworks in place as early as 2005. By 2010, all the Statsbygg projects had used openBIM principles like Industry Foundation Classes (IFC). Attention focuses on lifecycle costs and the environmental impact of infrastructure.

✔ **Russia:** BIM development is a fairly recent focus outside of academic institutions. In March 2015, the Expert Council, under the government of the Russian Federation, selected some pilot projects to explore the potential of BIM implementation.

✔ **Spain:** The first BIM Working Group in Spain has been set up by the government of Catalonia and Barcelona City Council.

✔ **Sweden:** The Swedish Road Authority, Trafikverket, is researching BIM for the Virtual Construction for Roads (V-Con). V-Con is a European project co-funded by the European Commission concerning standardization and implementation of BIM in the road construction and management sector.

The Americas

The following list gives a brief overview of the BIM activities taking place outside of the United States, across North, Central, and South America:

✔ **Brazil:** The National Department for Transport Infrastructure (DNIT) is embracing BIM, with major road schemes such as the 927-kilometer federal highway expected to adopt BIM.

✔ **Canada:** The Institute for BIM in Canada (IBC) has been formed by a number of partner organizations, including government representatives and buildingSMART Canada, to standardize the coordination of BIM across the Canadian construction industry. Canada BIM Council (CanBIM) has been a key driver for advocating BIM in the AEC industry.

✔ **Mexico:** The new international airport for Mexico City project is using BIM to produce a facility that will welcome 50 million passengers annually.

✔ **Panama:** The expansion project to the Panama Canal, one of the most important waterways with respect to handling global trade, has adopted BIM from the outset. MWH Global, a lock design specialist, is using BIM to construct a third set of locks to allow more traffic.

Middle East

Generally, you can see the macro-scale adoption in the Middle Eastern BIM as passive, and you can characterize BIM diffusion as *middle-out,* which means rather than large practices pushing from the top down, or smaller agile companies influencing quick acceleration, the core market is driving it. The Middle East has seen BIM demand in the market as being artificially created by the imported expertise of consultants from the United States and UK. Middle Eastern owners are requesting BIM, but the countries' organizational processes that should support that request aren't yet in place.

The following gives a brief overview of the BIM activities taking place across the Middle East:

- ✓ **Qatar:** Multiple projects are trialing BIM concepts. However, the development of a national or regional BIM policy should greatly help the success of Qatar's BIM mega-projects surrounding the FIFA 2022 World Cup and the achievement of goals such as the Qatar 2030 National Vision. BIM has huge potential to influence health and safety.

- ✓ **United Arab Emirates (UAE):** Dubai is the most populated city in the UAE, and Dubai Municipality has mandated the use of BIM for large-scale projects such as hospitals or stadiums, including buildings that are more than 40 stories high or 300,000 square feet. Interestingly, buildings delivered by foreign and international design and construction teams are automatically required to meet the BIM mandate.

Most clients in the Middle East aren't mature and are only starting to ask for BIM. Generally, no employer's information requirements (EIRs) are required, and the pull is mainly only from major clients and Tier 1 contractors.

Elsewhere in the world

A number of other countries are also insisting on the use of BIM, and countries like Malaysia and South Africa are at the very beginning of their BIM implementation strategies. The following list gives a brief overview of the BIM activities elsewhere in the world:

- ✓ **Australia and New Zealand:** The driver for BIM in Australia has been a long-overdue desire to get local territories and states to coordinate construction policy. Government and the Australasian Procurement and Construction Council engaged international help and buildingSMART chapters to develop a BIM implementation plan and supporting guidance. BIM is mandated on all public sector projects from 2015. To the mutual benefit of both countries, Australia and New Zealand are planning to work together to define the processes and protocols required for BIM, infrastructure, and land works. Recently, the extensive asset management plan for the Sydney Opera House and the ongoing work to rebuild Christchurch after the 2011 earthquake have both been demonstrated as exemplar implementations of BIM.

- ✓ **China:** National adoption of BIM is earmarked for 2016 and is being jointly organized through the Ministry of Housing and Ministry of Science and Technology, with a large influence from academia. However, the extreme pace of the industry to keep up with construction demand makes blanket implementation difficult. Like in many European countries, BIM is likely to be accelerated by infrastructure investment.

It's clear that national mandates for BIM need to be part of modernizing the construction industry, including bringing the environmental impact, accessibility, and structural safety of buildings to public attention.

✔ **Hong Kong:** The Housing Authority of Hong Kong requires BIM for all new projects. In tandem with the Construction Industry Council, Hong Kong is developing a BIM roadmap for successful implementation and is looking to make the most of operational data and standardized methods.

✔ **Japan:** BIM is an increasingly familiar subject, especially in Japanese academia and among professional institutes. The Japanese Ministry of Land, Infrastructure, Transport, and Tourism formed a construction task group made up of public and private construction industry members to develop BIM proposals.

✔ **Singapore:** Singapore is a swiftly moving market for BIM, and through its Building and Construction Authority (BCA) the government has embedded BIM into its building code checking and approval requirements. The e-submission process allows a project team to submit a federated BIM with prerequisite inclusions for planning and building permit regulations. More local agencies are adopting the system for approval, and it has extended beyond structural design to include services and environmental considerations. The BIM Roadmap, a construction strategy document, aims for 80 percent of the industry to be using BIM on all projects and, most importantly, to report on the return on investment.

✔ **South Korea:** The South Korean BIM Guide made BIM compulsory for all public sector projects by 2016, with a spotlight on long-term facility management.

Keeping Up to Date

The construction industry is typically adversarial and slow to change. However, to implement BIM and support digital ways of working requires a fundamental change in your approach. To prepare for new ways of working, the construction industry has needed to change and adapt a number of the aspects in which you operate. The following sections focus on why the UK industry has restructured existing plans of work and adopted a new unified classification system.

Restructuring plans of work

The Digital Plan of Work (DPoW) helps break down the barrier to BIM entry and build capacity. The deliverables from the DPoW, rather than the DPoW, are key.

A number of plans of work exist and each has a different focus. Plans of work have unique approaches to process. They also have wide variations in the amount of detail, depending upon the specific sector or domain's needs. For example, plans of work for buildings allow for design work to continue after construction has begun, whereas plans for work for infrastructure usually imply that all design is completed when the project goes out to tender.

The Construction Industry Council (CIC) has produced a new, coordinated UK plan of work involving the consultation of a wide range of institutions and organizations. The plan of work has eight clear project stages (from 0 to 7) encompassing the whole project lifecycle, from identifying strategic need through to operations and end of life. The naming of the project stages is agnostic with regard to project type and construction sector. The plan of work follows:

Stage 0: Strategy

Stage 1: Brief

Stage 2: Concept

Stage 3: Definition

Stage 4: Design

Stage 5: Build and Commission

Stage 6: Handover and Close-out

Stage 7: Operation and End of Life

A report undertaken by David Churcher and Mervyn Richards called "Cross-discipline design deliverables for BIM" aimed to define a single set of data drops during the design and construction phases across both building and civil engineering projects.

The RIBA plan of work has been the definitive model for building design and construction processes in the UK since 1963, and has a significant influence internationally. The RIBA plan of work has evolved over the years in response to changing processes within the industry, but has required its biggest overhaul since its launch, in response to the radical way BIM has altered the way the project team's process for designing asset design and development. Check out `www.ribaplanofwork.com/about/Concept.aspx` for a graph that shows the RIBA plan of work with its eight stages.

Going digital

A DPoW is a generic schedule of phases, roles, responsibilities, assets, and attributes, made available in a computable form. Produced on behalf of the UK government to support the BIM Level 2 process, the BIM Toolkit is specifically designed to enable the project leader to clearly define the team, responsibilities, and an information delivery plan for each stage of a project — who, what, when, where, and how — in terms of documents, geometry, and property sets.

Although the project team produces CAD or BIM information full size, the project team typically issues or exchanges it as drawings in hard copy (prints) or soft copy (electronic).

Standardizing classification structures

Adopting a classification structure is fundamental to BIM. *Classification systems* help project participants identify and locate things quickly. Think about a dictionary for a moment. This is perhaps one of the best examples of an effective classification system. It organizes information in a standardized way so that you can easily find and retrieve information. A dictionary also has a set of rules, or *taxonomy,* that are easy to understand and allow designers and other parties to add more information (and in the right place), and it also allows models to reference information and be better maintained.

These benefits underpin the purpose of all classification systems, such as Uniclass or Omniclass. The amount of construction information produced is vast and is generated by a number of people. Put into the equation the amount of information that capital programs can produce digitally and you can see that organizing information in a sensible and logical way is vital.

Classification structures such as Uniclass are being updated to take into account the requirements of object-oriented modeling. They've also been revised to enable a consistent approach to classification across building, infrastructure, and civil sectors.

Testing the requirements

The UK government has devised a simple ten-stage checklist to use when testing the requirements. Use Table 9-4 as a guide when looking to implement BIM. The checklist helps you determine whether the requirements have been achieved by asking a series of simple questions.

Table 9-4	Ten Tests of the UK Government's BIM Approach
Consideration	**Question**
Valuable	What are the cost savings against benchmark costs? Are there any improvements on key performance indicators (KPIs)?
Understandable	Does everyone understand the requirements? What communication methods are people using? Has a department implemented any training or educational needs?
General	What's the BIM adoption percentage of the department's portfolio value? Can the BIM approach be applied to all projects?
Nonproprietary	What percentage of portfolio value requires open data deliverables such as COBie and IFC?
Competitive	Is an information delivery plan (IDP) being used for procurement to maximize options?
Open	Are tools and file formats being prescribed in an open and agnostic manner?
Verifiable	Is data being tested and verified at agreed exchange points?
Self-funding	Does the BIM approach pay for itself? What about client cost and departmental funding?
Timescales	Is BIM Level 2 business as usual? What about the next phase?
Compliant	Do data proposals meet the employer's requirement?

Achieving the BIM targets

The UK government has set very clear targets around cost savings, carbon, and delivery times. Early adopter Level 2 BIM projects have regularly cemented cost savings between 12 to 20 percent against initial benchmark costs through BIM, and other initiatives such as soft landings. The Level 2 hypothesis is still to test the carbon agenda and embed it within the COBie data drops, but already trial projects are pointing toward better solutions.

Chapter 10

Standardizing the Modeling of BIM Objects

*P*rocuring data in any market is a complex task. Being clear about what you need and are prepared to pay for and then checking you have received what you expect needs careful planning. In this chapter, we introduce you to the digital building blocks that come together to form the models, more commonly known as BIM objects. Each level of complexity has more possibilities. Essentially speaking, a digital representation of a real-life construction product carries information not only about its physical appearance, such as its dimensions and shape, but also about its functional characteristics and performance.

As the construction industry builds more and more buildings and assets with BIM, the advent of clear data requirements brings many opportunities for value creation in the industry. The industry revolutionizes not only the way you produce construction information, but also how you can make best use of that information. To achieve this, the industry needs to standardize these digital building blocks that the project team uses to create digital buildings and assets.

You can discover that objects have unique geometry ranging from simple to complex, and they're available from a number of sources. The measurement of geometric data continues to be rather subjective during the journey toward a recognized digital plan of works, but with data definitions documented at the appropriate level, you'll be able to validate data deliveries

Moving from CAD block to BIM object

With traditional CAD, CAD blocks or CAD symbols represented products. Unintelligent vector lines were grouped together to make a 2D object. (You may even remember the Letraset stencils in the days before the computer ruled the office and the Rotring pen was king.)

CAD blocks provided some consistency in that you could pre-set their appearance; for example, line thickness or line color. You saved them to your library so that you (and the rest of the office or team) could use them again and again, rather than trying to reinvent the wheel every time. CAD blocks were great in that they saved time, because you didn't have to re-draw that door over and over.

The move from 2D CAD to 3D CAD meant designers could now start to coordinate spatial information. Some CAD blocks even had limited flexibility and intelligence, allowing you to dynamically change the geometry. For example,

if you had steel beams of different sizes, rather than starting from scratch, you took an existing steel beam CAD block and manipulated the geometry through custom properties, meaning you didn't have to create a block for every possible variation.

The shift from CAD to BIM fundamentally meant creating models rather than drawings. Construction professionals now not only think about what something looks like, but also what it is. Previously, a wall was a series of lines; now it's an intelligent 3D component, with properties and attributes about its performance, installation, and maintenance.

3D CAD objects provide huge benefits over traditional 2D CAD, but to fully consider the real benefits of BIM, you should coordinate the 3D object with information from other areas, such as the master specification.

electronically as they're delivered. This chapter looks at what information goes into or links to great objects, whether you're authoring them yourself or using an object library.

Figuring Out What BIM Objects Are

BIM objects are a digital representation and placeholder for graphical and nongraphical information about a real-life construction product. They're digital building blocks that the project team adds to a project model.

Think of an object as a container or repository of information about what the real life product is, what it does, and how it does it. It should contain all the information required to design, find, locate, specify, interrogate, and analyze the represented product.

Objects can take different forms. Like baking a cake, you make sure you get the key ingredients right to make a good object. In the next sections, we explore the essential ingredients for a BIM object and the appropriate levels of information needed to support the whole project timeline.

Recognizing key ingredients

So what key ingredients do you need to make good objects? An object is actually a combination of a number of things, including the following that help you answer the corresponding questions:

- **Information that defines the characteristics of the product:** What is the object called? How does the object perform? How is the object maintained?

- **Geometry that represents the product's physical appearance and characteristics:** How big is the object? What shape is the object?

- **Visualization data giving the object a recognizable appearance:** What does the object look like? What color is the object?

- **Functional data, such as detection or clearance zones or data that enables the object to be positioned or behave in the same way as the product itself:** Does the object open or close? Can the object only be positioned in a certain place?

Listing object types

Objects, sometimes referred to as *families,* are used throughout the project timeline, right from when you're creating shapes and forms at a concept stage through to making a product selection and in-use. They contain varying levels of information, depending on whether an actual real-life named manufacturer product has been selected. This product selection process occurs at different times, depending on the procurement route selected. The contractor, rather than the architect or engineer in the case of a design and build procurement route, may undertake it. Understanding the different BIM object categories is important so that you know what the object can be relied upon for. Broadly speaking, objects fit into three camps:

- **Generic object:** You use a *generic object* in stages of the design when you've yet to decide on the final project or solution.

- **Manufacturer object:** Also referred to as a *product object* or *proprietary object,* a *manufacturer object* is an object that reflects or represents an obtainable product provided by a manufacturer or supplier. This is when those who are tasked with providing as-built information add details such as maintenance and warranty data.

- **Template object:** You use a *template object* to guide the production of both generic and product objects by providing schedules of classification values and a minimum number set of attributes.

Now that you have considered whether your object is to take the form of a generic, manufacturer, or template object, you need to think about how best to model and represent your object. This can take a few forms, and objects are generally

- ✔ **Component objects:** These are building products that have fixed geometrical shapes; for example, doors, windows, sanitary-ware, furniture, and so on. You can further break down these objects into

  - **Static components:** An object available in one size.

  - **Parametric components:** An object available in a range of predetermined sizes, or the designer can determine the size.

- ✔ **Layered objects:** These are building products that don't have a fixed shape. Some examples are walls, floors, ceilings, and roofs. These objects are typically constructed from a number of layers (although just one layer is possible), and they don't have a fixed geometry. The designer's structural calculations define the geometry, which in turn determines a concrete floor layer thickness. Another example could be a wall, where the thickness of a wall is established; for example, it comprises a masonry outer leaf, insulation, and block-work inner leaf. However, the wall's height and length will be determined within the project environment. Manufacturers may also determine the thickness of the object layers; for example, an insulation board may be available in a set number of thicknesses.

- ✔ **Material objects:** They carry information regarding identity, performance, and appearance. They can be used on their own, as finishes and coatings, as building products within an object, or as a way to represent an option within an object. The level of information for a material to be qualified, quantified, and specified within a project environment varies from a simple name and description through to detailed technical information.

Just as real-life construction products have a number of options, such as color and size, BIM objects can reflect these variations and options from which the designer makes a selection. How this is best achieved depends on the individual BIM platform. A BIM object may be highly configurable, with parametric features and component (instance) properties that require the designer to make decisions, making it very flexible. For example, an object that could be fabricated in any size may have component (instance) properties for height and width. A BIM object with multiple options may be developed as individual objects that are then embedded into the overall BIM object, such as a window that can be made up of frame, mullions, and glass.

Some BIM platforms work by using a catalog file that loads multiple versions of a BIM object. This can be very efficient in presenting many possible product variations. You often handle variations selections through a text-based file.

The BIM platforms used by designers handle multiple-layered objects differently. Some platforms can model multiple-product-layered objects; others don't have this capability and the project team has to model the layers individually.

Typically, the project team delivers layered objects in a container file. This file can host the layered object and variations of the layered object, and the designer can select as required. For example, the manufacturer of a composite insulation board delivers all product variances in one container file.

You can use a number of BIM objects in combination to form an assembly. *Assemblies* are groups of components and spaces that allow many options to be described. By combining the objects in different arrangements, you can make many permutations of an assembly. For some construction products and arrangements, there can be millions of potential options.

Being aware of the timeline

Putting too much information into the model too soon may restrict both the design and supply chain options for others in the team and lead to other forms of waste with the inherent costs incurred by both the supplier and the client. It also places constraints on the supply chain to offer compliant alternative designs, offering better asset or value performance. Information requirements progress as the project develops. BIM objects develop with construction workflow and carry information that's relevant to the particular stage of work the information is intended for. Typically, an object starts life as a generic representation, and at an early stage within the design process it may just show basic 3D geometry. This information may be just enough for space planning, but at later stages you need information to manage the asset to its optimal performance.

As the design develops, depending upon the design selection and criteria, information within the object becomes more specific and detailed. The project team changes to a manufacturer object, containing the actual properties. Evolving from generic to proprietary objects during the design process — and not, as often happens, during the construction phase — means that users have the opportunity to see in real time the effects of the product selection upon criteria such as performance and lifecycle costing.

At the design stage, consider what information the project team requires. Include 3D information with specification information attached, so that architects and designers can undertake early massing studies. For more information about specification, check out Chapter 17. A minimum amount of detail should represent the space allocation for any given product's access space for maintenance, installation, and replacement. Also consider any operation space required, such as access requirements.

Obtaining BIM Objects through Object Libraries

An *object library* is a collection of reusable, predefined assets such as objects, materials, and textures that you can import into a BIM platform. The libraries provide a way to share and store knowledge and information. In fact, with so many product manufacturers across the globe and with each manufacturer potentially producing hundreds or thousands of project lines, ranges, and applications, object libraries are perhaps the only way to keep track of everything.

Contractors are currently developing libraries containing site-based objects including cranes, temporary works, and welfare facilities. These types of objects, together with common site services such as lighting, hoardings, barriers, and fencing, are useful in pre-construction health and safety planning.

In the following sections, we consider the different sources you can use to get your hands on BIM objects and how you can control and distribute these across your project team.

Describing object libraries

The client should document or reference specific controls and management of libraries in the employer's information requirements (EIR; refer to Chapter 8 for more information about the EIR). The client should also document details of model definition and model development.

The EIR describes the details of where the project team can access BIM objects. BIM objects are shared among the team either from a central client or project library or from an external source such as a public BIM portal. The EIR should also include aspects such as BIM object version control and the exact level of detail and information BIM objects should contain at various stages of the project.

Comparing BIM object libraries

You use a wide range of objects, not just from BIM object libraries or manufacturers' websites, but also content that manufacturers may have developed in-house or the standard out-of-the-box content that comes with your BIM authoring platform of choice. These sections identify the sources from which BIM objects originate and from where you can get your hands on them.

You can obtain objects in many ways, but be warned because they aren't all created equal. Make sure you know the source of the content and the standards to which they've been modeled. Just because something looks pretty doesn't mean that it's correct or contains the level of information you require.

Out-of-the-box software

Typically, most BIM platforms come with an out-of-the-box library of BIM objects that covers a broad range of objects that you may need when designing an asset, from building fabric such as walls, roads, and roofs to fixtures and fittings such as furniture and lighting. These objects are usually preloaded within your BIM authoring software and can be accessed through the BIM platform's user interface.

Out-of-the-box objects come with different degrees of parametric ability that you can then import into the project environment. These objects usually contain a set of basic properties such as classification. You don't need programming language or coding.

Public BIM library portals

In recent years, many third-party web-based BIM object libraries have emerged. These public library sites are usually free to the end user, such as the designer (although they may require a user registration process), but the sites charge manufacturers to create and host their content.

The quality, range, extent, and type of content of BIM objects vary from public library site to public library site. Some libraries may be vendor specific, and others have content in a range of vendor applications. Many libraries also have a nice community feel and are places to exchange, share, and upload your own content. However, the caveat here is that you don't always know the source of the content and the standards to which it's been authored.

Some BIM object libraries also offer a range of extensions or plug-ins that work with the BIM platform's interface to add additional functionality, such as dragging and dropping content directly from the library website into your model.

Ensure that it's made clear in protocol documents if you're planning to use vendor-specific proprietary objects on your project or aim for open BIM formats like IFC. The alternative is to ensure that all data can be exported into open formats from all the platforms being used.

Private BIM library portals

Project teams are using invitation-only, private BIM object libraries more and more to share content within a particular organization, peer group, or supply chain. You may use a private BIM library when your organization needs a high degree of security around the data that it shares.

Private BIM library portals usually carry a charge to the end user. This may be on an annual subscription basis or in some cases on a per-project basis. Exactly who picks up the tab depends on the individual project circum-stances. If the client is insisting you use a private library, then the project lead or main contractor may have to pay initially and then recover any additional costs within their fees.

Recognizing BIM Objects

BIM objects contain varying levels of sophistication, and graphical and parametric control. Graphical data may include information such as the size, shape, and area. To be able to interact, create, review, or manipulate BIM objects, you need to understand and identify the different types of information that they contain. A BIM object may contain the following information:

- **Dimensional:** This refers to the measurable extent of the object such as length, breadth, depth, or height. This information includes sizes, shapes, and areas associated with the graphical elements created within a project. Although a window frame may have many facets and extrusions, you must remember that you use BIM to implement a product, not manufacture it, and so to this end irrelevant information serves to add more data from which you're not getting any meaningful use.

- **Parametric:** This is the ability, using rule-based relationships between objects, to enable the project team to update related properties when one property changes. Instead of modeling a door over and over again, you can adjust it parametrically, thereby saving time. If a design decision is yet to be made, you can simply enter an arbitrary value and update it at a later date when you know the information.

- **Clearance:** This refers to the amount of space or distance the object requires, either for operation, maintenance, or health and safety requirements. Aside from the dimensions of the objects themselves, you need to consider their relationships to other objects and zones. An example is showing the clearance zone around plant machinery needed for maintenance, or the area required for a wheelchair to turn. This information and knowledge exists, and through BIM you can begin to add connections to this data.

✔ **Maintenance:** This refers to information that is required to preserve or keep the asset going in a good condition. The object should contain or reference all necessary information about the product. Then the project team can include it in handover documentation and attach it to any commissioning or handover documents so that the end user, client, or facilities manager can adequately maintain the asset. If you're tasked with providing as-built information, you should update information with any supplementary information, including maintenance records and replacement dates.

✔ **Connections:** This refers to the information required for an object to link, associate, or relate to another object. In an object-oriented world, objects have a relationship with other objects around them. Connections and associations — for example, a sink object with connections to services — greatly aid the designer when it comes to analysis.

✔ **Identity:** This refers to the information required to determine what the object is. Data must be categorized and arranged so that it can be easily retrieved; otherwise it's difficult to use. Standard formats drive the ability to use the data outside the BIM project file. So that other members of the project team can meaningfully reuse the information, make sure you provide a consistent set of parameters and attributes with consistent naming conventions. As a bare minimum, the project team should be able to identify a product by a trade name or model number. Ideally, the BIM object should also contain information like version number, issue date, and classification codes.

A description of a product's shape alone isn't sufficient to check whether it's correctly installed. Products and equipment may require surrounding operation space or have additional space requirements for transportation, installation, and assembly.

Deciding on how much is too much

Although you don't want to produce too little information, likewise you don't want to produce too much information. The key is providing the right information that is relevant and that is provided at the right time. Spend a moment to decide and plan how much time and effort you spend in providing information. Too much detail can reduce productivity and create waste during the design process because large models require more processing and longer file transfer times. On a large project the model can become unmanageable and may need to be subdivided.

Take a moment to think about whom the information is for and how it's going to be used. For example, the detailed modeling of a toilet lever flush handle may include many facets and extrusions. However, at an early stage of design a generic object has little need for very high levels of detail. The project team or client may not have decided on the final toilet and flush handle.

The information overload will be exacerbated if more is included than is needed. You need to provide an appropriate amount of content at each stage. Although you can easily get carried away when modeling information, only model what is required. The amount of data should respond to the original questions that the client has asked in the EIR. You may see these questions referred to as plain language questions (PLQs).

In some instances, 3D geometry may not be required or appropriate. An example is in modeling a metal window. You model the outer profile of the frame but not the intricate internal framing members — you may represent these using 2D line work incorporated into the object and in digital 2D outputs of the project.

Consider parts of objects that you won't model. Examples are fixings such as adhesives, screws, and bolts. They're too small to model because too much detail compromises the usability of the object.

If in doubt, leave it out! Delivering a greater level of detail than you need is wasteful to the supply chain and may overload the IT systems and networks available. This problem is reduced as available processing power increases, either through internal IT or moving model tasks into the cloud.

Categorizing level of development

The US American Institute of Architects (AIA) document E202 defines level of development for model elements. Think of *level of development* as the overarching requirement for both the graphical level of the object's geometry, which is called the level of detail (LOD) and the nongraphical data in the object, which is called the level of information (LOI). These two levels in combination form the level of development.

Note that you might also see the acronym LOD being used for the overall terms level of definition or level of development, especially in the United States. This can get confusing when you're being asked to indicate what "LOD" you'll be delivering. Make sure that everyone is clear on what any acronyms mean and that everyone is speaking the same language. Use our glossary to find simple definitions you can use on projects.

Because of the recent revolution in digital working processes and BIM in the United States the AIA decided to cut short its once-per-decade update process and release a set of dedicated contract documents about BIM and digital data. These provide frameworks and protocols for integrated project delivery and information exchange, and are as follows:

- ✔ **AIA Document C106, Digital Data Licensing Agreement:** This agreement clears up the legal liability and copyright permissions associated with collaborative project teams — for example, that any party exchanging data grants a limited license to the receiver to use the information on the project and guaranteeing ownership rights.

- ✔ **AIA Document E203–2013, Building Information Modeling and Digital Data Exhibit:** This document is attached to existing contracts, such as the Standard Form of Agreements between owners and project parties. You can consider it similar to the UK PAS 1192 documents, in that it establishes expectations for data exchange and provides guidance on developing a detailed process protocol for BIM and digital data, as a statement of intent.

- ✔ **AIA Document G201–2013, Project Digital Data Protocol Form:** In connection with Document E203, this G201–2013 Protocol Form documents the procedures for digital data communications and submittals — for example, the use of electronic data management systems (EDMS) and where data will be stored.

- ✔ **AIA Document G202–2013, Project Building Information Modeling Protocol Form:** In connection with E203 again, this document is specifically focused on BIM and the management of model information. In particular, this is where the level of development descriptions is defined so that everyone can agree on what the level means.

It's worth noting that the AIA documents draw on a lot of previous information in earlier contract documents. The updated set includes content from the former AIA E201 and E202 releases. If you're looking at this area of BIM in detail, refer back to these original documents, but some are only available via AIA software.

If you want more information on the AIA Digital Practice Documents, a great guide is available at www.aia.org/groups/aia/documents/pdf/ aiab095711.pdf.

Defining the model's complexity

AIA G202 talks about five categories of level of development (100, 200, 300, 400, and 500). Table 10-1 shows the potential levels of detail (LOD) and levels of information (LOI) for an example BIM object, in this case a boiler.

Each number indicates an increase in the amount of graphical and nongraphical data across the project timeline. As levels increase, the data is less generalized and becomes more accurate and more product-specific.

Table 10-1		Levels of Detail and Levels of Information	
Level of Detail (US)	*Level of Detail (UK)*	*Geometric Detail*	*Level of Information*
100	1	Approximate dimensions and concept graphics for visualizations.	The project requires a boiler. Initial options described.
200	2	Generic boiler system, space arrangement, clearance zones, and input/output.	System description and initial requirements specified for performance of boiler.
300	3	Development of detail, dimensional constraints for technical design, materials, and finishes.	System selection and detailed performance requirements with reference to standards.
400	4	Technical design proposals, detailed spatial coordination of related systems.	Technical system specification, including components to permit product selection.
500	5	As-built installation and coordination, future use as maintenance information.	Detailed specification of manufacturer's product, including testing, operation, and maintenance.

Eyeing the levels of detail (LOD)

Vico Software first used the term *level of detail,* but various AIA committees have developed the framework and concept over the years. Under the guidance of the BIMForum, which is a unified, cross-industry group representing ten discipline sectors that aims to encourage BIM adoption in the AEC industry, the Level of Development Specification was launched in 2003. The Level of Development Specification has received further updates and the latest draft release is available for download at `https://bimforum.org/lod`.

Understanding the Model Progression Specification (MPS)

We talk about the BIM execution plan (BEP) in Chapter 8, but BEPs don't always pick up on the project-specific aspects of digital information and try to generalize the BIM process at a high level. Model Progression Specifications (MPS) are intended to explain how a project model will be developed over time and used by cost estimators, contractors, and other members of the project team.

It's important to understand that you'll probably create separate MPS templates for different projects. The way you coordinate and exchange data for a private housing project will be different from the plans for a sports stadium or an airport.

Recognizing the Model Element Author (MEA)

You hear lots of different names for the developers of model content, but the AIA documents use a specific one, the Model Element Author (MEA). This assists with the BIM audit trail, because one individual is assigned the role of developing a model element to a pre-set level of development within a certain project stage.

The indication of MEAs feeds into the overall BIM framework, very like the responsibility matrix that you can find in the UK RIBA Plan of Work at www.ribaplanofwork.com.

BIM objects usually contain information to help the contractor to purchase the construction product, not to fabricate it or manufacture it, although some objects can be developed further to include this kind of fabrication data. Therefore, objects should have an appropriate level of detail and shouldn't include more detail than is required or is useful.

Renderings are photorealistic outputs from the BIM that show a more accurate depiction of the material than in the model view. BIM platforms allow for the configuration of the rendered appearance of the materials, such as transparency and reflectivity.

You can also use image files to represent the material's appearance. Image files such as bitmaps and bump maps can give an additional appearance of texture to image files. You need to scale the image correctly and allow for a repeated pattern.

Embedding data into the objects

Not all the data lives in the model. It becomes impractical if the objects and hence the models are overloaded. The linking to other relational bases provides a wider source of information and empowers the object, making it a rich source of information. Think of a BIM object as a placeholder — not only a physical representation of the real-life physical properties of the said object, but also a home for nongraphical information such as performance criteria, cost, and operational details.

The next sections examine how to handle BIM objects that have multiple variations and options and consider the most appropriate place for information, either in the geometric model or with nongraphical data such as words and numbers that are usually found in associated documentation like specifications.

Evolving the level of information (LOI)

In contrast to the LOD, which is purely a geometric and 3D visual requirement, the LOI is about the amount of accurate nongraphical information in the object.

The most obvious way to notice LOI is the amount of properties or attributes you can read when you open up an object in your chosen BIM platform, such as performance criteria like fire resistance or thermal transmittance or warranty data like replacement cost or date of installation. The key difference is that most of the nongraphical information can be described in written forms.

An object can be highly graphical and include detailed rendering options but contain very little property information. Some information is more appropriately located in the *geometrical* part of the BIM object, such as the physical size and shape of the 3D model, whereas other information is more suited to the *properties* part where information is described in a nongraphical way. The content of a project specification can be linked to many of the properties in the object, instead of the data being duplicated. The specification is part of the project BIM, and modern specification tools model their clauses as objects to create relational databases and intelligent linkage between models and drawings and the written contract documentation.

Author it once, and in the right place; report it many times

Information comes from a variety of sources and BIM tools (for example, BIM authoring platforms as well as cost-estimation software and the specification). BIM objects have properties, and most also have geometries (although some don't; for example, a paint finish). To avoid duplication, information should be both structured and coordinated. In traditional documentation, you want to author it once and in the right place.

With BIM you want to be able to author the information once, in the right place, so you can then report it many times. In other words have information in just one place to avoid duplication and errors and in a form and format that can be configured and viewed in many different ways, by many different people, for many different applications. For example, the structural engineer may only be interested in viewing structural load-bearing walls and not non-load-bearing, or the supply chain may use information about an object's weight to determine shipping and delivery charges.

Mind the gap! Don't underestimate the power of words

At a certain point in the project cycle, the written word can take you to a deeper level of information. Within a textual context, you describe the length, height, and depth of something. Words help you describe the project specifics and the workmanship.

For example, consider an analogy of a BIM object representing a simple cavity wall. When you model it in your chosen BIM platform, you define the height, width, and depth. You may also define the outer leaf, inner leaf, and insulation layers. However, when you compare the level detail within the wall modeled in your BIM platform to within the specification, you can begin to appreciate the level of detail that's missing.

Information contained within the geometric model isn't always relevant to the specification; for example, the height of a wall. Information within the specification sometimes has no bearing on the model; for example, workmanship or execution instructions. You don't want to duplicate information; therefore, what you really need to strive for is a two-way association between the spec and model to ensure consistency and enrich the objects.

You don't need to put all information within the object. Sometimes linking to an external source (maybe a manufacturer's website) is better. For example, a hyperlink to a PDF document is a valid attribute.

Supporting Standardization

Standardization is an important consideration because it provides a common benchmark for quality and assurance that the construction industry can work to achieve. It provides clarity and a way in which you can assess products and services and either accept or reject them. The construction industry working toward the same standards creates a competitive environment, and it also allows global trade to occur because the construction industry isn't working in silos.

Comparing TVs is fairly simple in an electronics retailer. At a quick glance they may all look the same — the typical properties, information, and primary considerations that a consumer makes before purchasing have all been standardized. The retailer has captured key information about the product so that the consumer can make an informed decision before parting with her hard-earned cash (or credit card!). All the pertinent information is clearly available.

Unfortunately, the same can't always be said for product manufacturers in the construction sector. That's why when creating a BIM object you need to think carefully about how the real-life construction product the BIM object represents will be used and the information that's included. In the world of BIM, the UK BIM Task Group has published a variety of standards, including BS 1192-4 COBie Standard and PAS 1192-2 (www.bimtaskgroup.org). These encourage standardization and are focused on the production, exchange, and use of information as the means of delivering improved performance across the whole life of a building. Vast amounts of information are created during the construction phase, but much is lost or wasted. You need to safeguard against information loss and start managing and analyzing information digitally. Remember that BIM isn't architecture; it's data management.

Following a consistent and standardized approach to creating your BIM content is a must. In the next sections we consider common approaches to modeling and how to be consistent with your property sets.

Following common approaches: Modeling requirements

To provide more efficient and accessible BIM objects, you must take a standardized approach. Creating digital buildings requires a consistent kit of parts that can yield all the benefits that standardization brings. Objects are easy to source and to use, and are comparable, interoperable. By standardizing the information within objects, you can compare them and make an appropriate selection for the project.

Common approaches to the modeling of the physical characteristics of products make the BIM objects simple to use, affording the designer a reliable, consistent, and intuitive experience. The hard work is in the detail; for example, BIM objects in IFC format. These IFC files are manipulated so that their information properties are consistently grouped and organized, which makes their use in various BIM software straightforward and consistent.

Another example is the use of standardized properties. The benefits become obvious when using objects from more than one manufacturer in the same project. When creating schedules that span products from many manufacturers, using a standardized property enables you to display information relating to each of these products in a single column, much in the same way you find the number of megapixels listed when comparing cameras.

You can find a lot of discussion online about object properties and naming conventions, but there isn't much agreement on standardization. Some organizations are working to help the situation. For example, the NBS BIM Object Standard (www.nationalbimlibrary.com/nbs-bim-object-standard) defines minimum standards for BIM objects. It refers heavily to the BS 8541 series, a code of practice that takes the form of guidance and recommendations for library objects for the AEC industry, as well as COBie and the buildingSMART IFC schema.

Being consistent: Standard property sets

Consistency is fundamental to being able to correctly export information from within the object and use it in other applications. To meaningfully reuse the information, it must have a consistent set of parameters and attributes with consistent naming conventions. As a bare minimum, you can identify a product by a trade name or model number. At an early stage within the project this type of detailed information is unlikely to be known, as specific product selections are yet to be made. This is where a classification system such as Uniclass or Omniclass comes into force, as a way of the whole supply chain knowing that everyone is referring to the same thing. You may call a spade "a spade," but your contractor may call it "a shovel" and your client "earth relocation equipment."

A *parameter* is a property or attribute of an element; for example, fire resistance, material, type, and color. A *property set* provides a consistent set of properties across all objects, giving information such as version number, issue date, and classification.

Include standardized construction data and recognized standards such as

- ✔ **COBie properties:** This data standard was developed by the US Corps of Engineering to manage the data coming from BIM models into the client organization, particularly for the handover of operational and maintenance information.

- ✔ **Classification:** *Classification systems* are the tools that determine groups of things based on similar characteristics. They arrange information about a particular topic, coordinate and disseminate information, find and understand information, and join like information together. Whether you're in the supermarket or searching for information on Google Maps, classification is the key to finding what you want.

- ✔ **Interoperability properties:** To exchange information about a building using common and understood rules, you use the Industry Foundation Classes (IFC) specification developed and maintained by buildingSMART International as its data standard and registered with ISO as ISO 16739.

Table 10-2 lists some essential properties you should include within your BIM object and what questions these properties are trying to answer.

Table 10-2	BIM Object Properties
Property Name	*What Do I Answer?*
Author	Who created the object?
Name	What is the object called?
Description	What is the object?
Version	How up to date is the object?
Revision	Has the object been updated or modified?
Globally Unique Identifier (GUID)	How can you identify the object?
Classification	What is the object?
Performance data	How good is the object?
Quantity	How many of the objects are there?
Manufacture, model, and serial number	Who makes the object?
Position	Where is the object?
Operating instructions	How does the object work?
Maintenance instructions	How is the object cared for?
Fault-finding instructions	What happens when the object goes wrong?
Commission instructions	How is the object commissioned?
Health and safety	Is the object safe?
Statuary testing	When was the object tested? When does the object next need to be tested?

Enabling Interoperability

You collate, produce, submit, and retrieve information digitally. The benefit is that you can manipulate digital information to suit different contexts, requirements, and exchanges. However, in a construction context you haven't always been very good at exchanging this information, particularly when using a number of different software solutions.

Within a vendor's own suite of software, things are usually pretty good. In everyday life, this may be embedding a Microsoft Excel spreadsheet into a Microsoft Word document, or receiving an email that contains a postcode/zip code using Gmail and opening the exact location using Google Maps. However, you can't complete a construction with just one piece of software. The software from different vendors needs to communicate with each other, and then the project team needs to agree on communication rules.

One method is to consider interoperability through open standards. Simply put, *interoperability* is the collaboration, exchange, and ability to use and interact with BIM data between different BIM platforms. Open standards are commonplace within computing and normally people take them for granted. For example, sending an email from Microsoft Outlook to Google Gmail is possible through the data standards RFC 5322 and Simple Mail Transfer Protocol (SMTP).

With the construction industry, IFC and COBie are ways in which the whole project team and supply chain can communicate and exchange information with each other, regardless of the software they have used to produce the information. In the next sections you look at these two mechanisms to transfer information back and forth.

Industry Foundation Classes

BIM authoring applications generally allow the user to identify a model subset (or filter) when exporting to IFC by exporting only the layers that are currently visible in the BIM-authoring application and allowing users to export only the parts of the model relevant to the purpose of the export.

In order to do this, the project team must categorize objects correctly. Some BIM platforms automatically assign IFC information based upon the IFC schema. Others require additional properties (for example, in Autodesk Revit the properties IfcExportAS and IfcExportType are used). Generally, software solutions don't support the entirety of a schema such as IFC; they support an industry-relevant subset that's generally termed a model view definition (MVD). Software may be certified in terms of how well it supports a view definition. That is, a *view definition* provides a relationship between the whole schema and the software solution that implements it.

COBie

When we refer to COBie, we're referring to Construction Operations Building information exchange. COBie is a model view definition (MVD) of IFC that is a spreadsheet mapping of the FM handover view definition. This means that COBie is only concerned with a particular snapshot of the IFC schema that specifically looks at information that is useful to facilities management. COBie defines how the project team is to structure information and what the minimum data fields are. It's a data format, not a standard on what information the project team is to provide for facilities management. COBie isn't a predefined list of what information the client requires. So to ask for COBie without defining what you want in COBie is a bit like asking for a MS Word document without saying what you want written in it.

You can produce information for COBie in a number of ways. These include, but are not limited to, the following:

- Direct creation from the BIM platform. Objects may contain the relevant COBie properties.

- Production from an IFC file, using IFC-to-COBie translation tools and settings, based on properties as defined by buildingSMART IFC2x3 basic FM handover view.

Whatever method you choose, make sure that the properties are consistent and don't contain a hybrid of the two. You can extract a COBie deliverable from the IFC if the data is structured and exists within the file. Likewise, you can push COBie information into an IFC file if the information is structured and exists within the file.

The total COBie deliverable is provided by a range of people and comes from many sources, and you can't populate all information from the BIM platform. The BIM object should include COBie properties that don't include parametric behavior, graphical, or stylistic information.

Value associated with certain properties depends on the stage of the project. Take, for example, the property "InstallationDate." No value is associated because you don't know this information until the product is installed. You complete some values such as "AssetIdentifier" at handover stage, where the asset is made available for use or occupation. In the case of generic objects, you don't know properties such as "Manufacturer," and therefore you enter a default value of "n/a" (not applicable or not available). This acts as an aide-memoir for information that may be completed later on in the project, often by a different person or organization.

For now, the UK government requires for BIM projects only COBie, native model files, and reviewable 2D PDFs. You can deliver COBie as a spreadsheet or an *XML file* (a structured text file, a bit like HTML). XML files require software to make them readable for people, so unless a software (or web service) is proscribed, the client normally requests COBie as a spreadsheet. The theory is that people can fill in a spreadsheet manually, yet you still retain the possibility of populating data automatically via software from BIM.

The COBie file, which can be as simple as an Excel spreadsheet, holds valuable information for the asset management. The data is exchanged using spreadsheets to keep the complexity of systems and training to a minimum. It comprises 16 tabs that hold information, from very generic things to very specific things, and lists all the products installed in a building including manufacturer details, replacement costs, warranty information, and links to maintenance instructions.

Don't think you could deliver COBie through entirely manually filling in a spreadsheet. COBie spreadsheets may be readable for people, but they're not human friendly (except maybe to computer programmers). True, you could set up human-friendly spreadsheets that fed data to COBie spreadsheets, but that wouldn't overcome the enormous amount of data that you'd have to manually type in, check, and verify.

Think of COBie as a filter. COBie is a bus that gets you from one place to another. Or, to think of it another way, it's a bag with various compartments to put your data into. The number of the COBie worksheets to fill in depends on the project stage. Project team members only enter the data for which they're responsible. Designers provide spaces and equipment locations. Contractors provide manufacturer information and installed product data. Commissioning agents provide warranties, parts, and maintenance information.

COBie isn't just going to fall out of your BIM. You need time to set it up, test it, and then do the export and validate it. Keep in mind that it will take a lot less time if you use someone who knows what she's doing. Ideally, the information manager should program COBie data drops to avoid other deliverables. Expecting COBie to be delivered on the same day as a milestone document issue is silly. You need them for different purposes, so why increase everyone's stress levels?

Chapter 11

Collaborating through BIM Requirements

The first time you open up websites or books that try to explain BIM, they almost always end up referring to complicated standards documents. The bunch of impenetrable letters and numbers can look downright confusing. PAS what? ISO what now?

Yes, we have to refer to standards in this chapter, because they're the glue that holds collaborative BIM together and helps the industry to be consistent in its approach to this complete change of practice. We don't expect you to have read the standards cover to cover, and we know you won't understand them straight away or even be able to remember the difference between them all.

This chapter is a gradual, gentle introduction to a wealth of key documents and associated methods that aim to steer the processes and procedures of a notoriously complex industry toward collaborative harmony. Now we admit that this chapter focuses on UK BIM, partly because that's where our expertise is, but also because we think the suite of documents supporting "Digital Built Britain" does a lot of things right.

Think of this chapter as a quick reference guide to the documentation you should read eventually, after you're getting started with your own BIM implementation plan. BIM documents are like the Complete Works of Shakespeare; this chapter provides you with just what you need to know to get you through.

Running through Open Standards

On any given project, you want to be able to develop the ability to work effectively and efficiently no matter how many external consultants or teams are involved in the job. You should build up to pre-planned coordination and management of all the various parties in the project team through BIM standards and protocols. The full *buy-in* of everyone who'll access the information model is fundamental to BIM success. Everyone needs to be on the same page . . . of the same BIM standard.

In these sections we explore why standards are vital to the BIM process and how they can form part of your best practice approach to collaboration.

Grasping why standards are important

Today's world is built on standards that help people drive innovation and increase productivity. On a more rudimentary level, standardization makes your life easier; think about the standardized units of measurement that you encounter every day or labeled sizes that let you easily choose clothes or shoes.

Think about food labeling, which helps you to clearly understand the nutritional value of food and compare like for like, whether it's the sodium content of lasagna or the amount of protein in an energy bar. Standards exist where allowing the providers of a service or the manufacturers of an object free reign would result in confusion or chaos.

Start by thinking about what would happen if you didn't have access to standards:

- **Products of poor quality:** Items that don't meet standards may malfunction, break down, or even be unsafe.

- **Products that are totally incompatible with others:** Imagine if every electrical item in your home had a different proprietary plug or socket design, or if every model of car needed custom tires.

- **Every manufacturer's solution would be unique:** The options to compare standard components or their performance and to learn from experience and previous mistakes wouldn't exist.

Now just replace the references to products and manufacturers with projects and design teams. If you didn't have access to standards, you'd have

- ✔ **Project work of poor quality:** Construction processes that don't meet standards may malfunction, break down, or even be unsafe.

- ✔ **Project work that's totally incompatible with others:** Imagine if every drawing or model in your project used a different proprietary file format, and if you had to buy different software for every single team.

- ✔ **Every designer's solution would be unique:** The options to compare standard components or their performance and to learn from experience and previous mistakes wouldn't exist.

Interesting, yes? As you can see, BIM standards are vital.

In previous chapters, we discuss examples of standardization to demonstrate how valuable it can be — from credit cards to aeronautical design. We discuss in this chapter that the construction industry is ready for standardization and look at the documents that aim to equip the industry for a successful collaborative future.

Training the industry to think about open BIM

In today's digitized world, standards allow people to create, use, and maintain information in a well-organized way. The standards not only encourage best practice, but they also offer a means to achieve real and measurable improvements. Sharing construction information, drawings, specifications, and schedules in an agreed and consistent manner can bring about savings in cost and reduce waste.

In order to achieve this, the members of the project team must be able to work together more effectively than ever before. The following processes are critical to the modern BIM-enabled project team:

- ✔ Coordinating project information so that everyone has the most up-to-date version of the accurate and specific information they require for their next task.

- ✔ Communicating at the earliest opportunity; for example, when a member of the project team sees an issue with constructability, highlighting the problem before it reaches site and before it impacts other design work carried out in parallel.

- ✔ Reducing duplication of effort or abortive work by communicating decisions or analysis to the whole project team.

In the UK, the government has a clear BIM strategy that focuses on the efficient production, exchange, and use of data and information as the main means of delivering improved construction performance and savings.

The construction industry needs to safeguard against information loss and to start collecting, producing, submitting, and retrieving information digitally. You can currently manipulate digital information manually to suit different contexts, requirements, and exchanges. The next step is to enable automated information exchange, improve analysis capabilities, and encourage compliance checking.

In this emerging BIM environment, content needs to be

- Open
- Accessible
- Structured
- Understood
- Controlled
- Secure
- Standardized

To achieve this, you need to have standards around data, processes, and terms. Consider:

- **Where is your information stored and backed up?** At some point in the life of the asset, components will require replacement, upgrade, and maintenance. Knowing where to find the relevant information about the component at hand in the first place will save time and frustration.

- **Who owns the data?** The specifics of this question will be covered within the contractual documents. Make sure everyone within the project team including the client is clear about where the ownerships rest.

- **What does the data represent — is the project sensitive or confidential?** For example, making all information about a secure prison available wouldn't be wise. It's not unheard of for prisoners to get hold of sensitive building plans and tunnel their way out.

- **Who has access to the information?** Although not all team members need access to every piece of information, they do need the right information to contribute to a successful collaborative environment.

✔ **How can you control access — can everyone manipulate it?** Not everyone requires read-and-write access. In many cases, a read-only version of the data is adequate.

✔ **How do you record changes? Can you track who's accessed the information?** Having a robust audit trail helps you see who has accessed and changed information, which may have an effect on others.

The opposite of *interoperable* (or open) is *proprietary* (or closed). Consultants and construction professionals sometimes use the word *native* rather than *proprietary*. If the client asks for native BIM outputs, you need to include the original file format of your 3D information as well as the 2D deliverables that you've extracted from the model.

A proprietary data format that's particular to a BIM software vendor can be quickly, reliably, and efficiently updated and adapted to suit a changing market. However, in the long term, proprietary data formats prove expensive to maintain and support. You want to be able to exchange data no matter which software it came from; this is the very ethos of BIM.

Standards for the open exchange of digital data aren't a new thing. They started to emerge in the late 1970s, based on agreements between the leading CAD vendors and users. In fact, in the mid-1980s the International Standards Organization (ISO) technical committee (TC) 184 Automation systems and integration, subcommittee 184/SC 4 Industrial data, developed something called the Standard for Exchange of Product (STEP) model data. ISO created STEP (ISO 10303) because ISO thought that none of the existing formats, on their own, could support an open standard across multiple industries.

The problem was that STEP took a long time and was just too slow and unresponsive to meet the fast-paced needs of an upcoming market in the construction industry. You can imagine that motivation started to develop for something specific to the architecture, engineering, and construction (AEC) and facilities management (FM) industries, and a separate way of working began to evolve from the structure of STEP — open BIM and the use of Industry Foundation Classes (IFC).

You rarely hear anyone use the full term "Industry Foundation Classes." BIM users commonly understand IFC without necessarily knowing what the acronym means, just like people say JPEG or Scuba (joint photographic experts group and self-contained underwater breathing apparatus, in case you're interested . . .).

Standards 101 Betamax versus VHS

One of the most common analogies you hear about standards and the early adoption of BIM software is Betamax versus VHS videotape formats. The risk of early adoption (and high levels of investment in software and training) is that you may be backing Betamax, when everyone's eventually going to use VHS. The conclusion a lot of people arrive at is: "Shouldn't I just wait for a clear 'winner' to emerge?"

The analogy is flawed for a few reasons:

✔ JVC's Video Home System won the battle of *consumer* videotapes, but Betamax's sister Sony Betacam became the industry standard in TV production for decades. How do you perceive the tools your business needs to thrive — a ubiquitous home video market or high-quality TV production?

✔ Lots of examples exist of industries where no single market leader has a monopoly, and in fact competition is good for consumers. The key is standardization and *interoperability,* which means the ability to exchange, view, and operate the same format regardless of the application you're using. JVC persuaded Panasonic, Hitachi, Mitsubishi, and Sharp to adopt the VHS standard format for its VCR machines without licensing it. In fact, JVC and Sony had collaborated together on the pre-cursor U-matic format, but had different ideas about how to innovate.

✔ Until the eventual mass acceptance of DVD formats, VHS dominated the video recording sector for almost 20 years, but the format itself received very little development. A VHS cassette from the mid-1990s is identical to one from 1980. No one wants the same to happen to BIM software platforms. You want the tools you use to evolve and develop and not be held back by the need to work with other tools.

Heading in the right direction: open BIM

Open BIM is an initiative of buildingSMART and several leading software venders and is a universal approach not only to collaborative design but also to the delivery, operation, and maintenance of assets, which are based upon open standards and workflows.

Open BIM is more than just IFC. It's a commitment to open standards and engagement with the whole team across the life of a project. As a schema, IFC itself doesn't and can't provide interoperability alone; rather, it relies on the software packages interfacing with it. The schema sometimes sparks debate and criticism for omitted data or lost geometry, but is this due to the IFC standard or how the IFC schema is implemented in a particular platform? Limitations currently exist around IFC's ability to contain parametric information and manipulate the size of an object; however, IFC Release 4 (IFC4) and subsequent future releases look to address this.

Today, most modern BIM authoring platforms support the import and/or export of IFC model data. buildingSMART even issues official certification to applications that comply with consistent procedures. This flow of information is critical for collaboration and interoperability, because it allows use between different authoring and downstream applications — just think of when facilities managers or structural engineers use your information.

buildingSMART together with a number of leading software venders designed open BIM around some fundamental principles to encourage the benefits of interoperability:

- ✔ No matter what software various teams are using, you should be able to participate and collaborate equally. buildingSMART calls this a "transparent, open workflow."

- ✔ The size of BIM platform vendors isn't an issue, because both small and large software companies should be able participate with system-independent solutions. They use an interesting term for this, "best-of-breed" — aiming to promote competition and innovation but within a specific area.

- ✔ The aim is that the construction industry can use a common language throughout client and commercial groups during the procurement process, making comparing and evaluating outputs, service levels, and the quality of data much easier no matter what tools the project team uses. Software versions and upgrades shouldn't cause problems in your project either, because they all should be interoperable.

- ✔ By using open formats and standardization, open BIM also hopes that the project team will be able to use the output project data for much longer. If asset lifecycle tools for facility management can use the same formats then this avoids duplication of input and the errors that could result.

In the past, interdisciplinary collaboration has required cross-referencing each other's 2D drawings, requiring the project lead to manage coordination manually. Clearly, complex 3D elements need a more robust level of coordination, and so open BIM uses a concept of a reference model. One of the benefits here is that open BIM maintains your authorship.

If you're a consultant or supplier, because all vendors can participate in open BIM even with competing products, you can join any open project workflow without giving up the BIM tools you're used to. If all projects were open, then nobody should be excluded based on their software platforms. You can work with anyone without having to use the same platforms.

Steering the Industry toward IFC

This open approach is as much about a new mind-set as anything else, because it requires a strict regime of classifying elements correctly, in order that all actors can filter and exchange all information. Doing so is going to require champions of this way of working (like you, hopefully) to encourage and steer the industry towards interoperability.

In the following sections, we look at not only IFC, but also the other open standards within the buildingSMART family and why you should consider using them.

IFC: Building bridges

BIM is often a solution to a problem that you didn't know you had. However, parts of BIM can seem like solutions to problems you don't have, not in your particular business anyway. Understand what part your information plays in the wider project team and find methods that meet your specific importing, editing, and exporting needs. IFC is an industry-wide, open, and neutral data format that's fast becoming the de-facto standard for rich data exchange.

Ask yourself the following questions about the data you'll use:

- ✔ What data are you receiving from others? What format is it in?
- ✔ What data are you supplying in proprietary formats?
- ✔ What other outputs and published documents do the client or the design team expect as project deliverables?
- ✔ What file formats will the rest of the project team be using?

If the people you're working with are all using the same software, then interoperability isn't an issue. But in the vast majority of projects, someone will be using tools you don't use. Especially on international projects that move beyond traditional architectural, engineering, and construction (AEC) and into civil engineering, you may see structural modeling or data analysis that you haven't even encountered.

More commonly, projects may use two of the four big brands of BIM platforms: Autodesk, Trimble (including Tekla), Bentley, or Nemetschek (including Graphisoft), and you want to work together. That's where IFC aims to build the bridge.

Meeting buildingSMART International

So who's behind IFC? Well, the Industry Alliance for Interoperability was formed in 1994, involving 12 companies under the leadership of Autodesk and HOK, to prove the concept of interoperability and full information exchange between various software used across the building industry.

As part of an effort to develop a nonproprietary standard and take it global, the Alliance renamed itself the International Alliance for Interoperability in 1997 and it became a nonprofit group. The Alliance's aim was to support the creation of open, international standards for data exchange, moving away from private and proprietary standards that were a bit locked down.

Renamed buildingSMART in 2005, the organization comprises chapters that are national membership organizations sharing the vision and goals of buildingSMART International. Chapters also develop and promote the use of open BIM in their countries through education and publications.

Investigating IFC

You may have seen people use the term IFC a lot when talking about interoperability and sharing information. In simple terms, IFC schemes provide the guidelines to determine what information is exchanged about an asset. Think of them as the rules for sharing the right data, a fundamental part of BIM.

IFC exists to help you collaborate by generating the rules, the *model specification,* as a standard. IFC schemes may include geometry, but aren't limited to this. IFC presents your tangible building components, like walls and doors, and it allows you to link alphanumeric information (properties, quantities, and the classification structure) to your objects and maintain those relationships.

Since 1996, six principal schema releases have occurred: IFC1.5.1, IFC2.0, IFC2x, IFC2x2, IFC 2x3, and IFC4. (IFC4 is formally known as IFC 2x4.)

IFC4 is now registered with ISO as an official International Standard (ISO 16739:2013). buildingSMART hopes that software vendor implementation will increase. For further information on currently certified software and the software certification scheme, see the buildingSMART website (www.buildingsmart.com).

Supporting buildingSMART standards

Five basic standards and supporting frameworks overseen by buildingSMART surround open BIM and interoperability. They are

- ✔ Industry Foundation Classes (IFC)
- ✔ Information Delivery Manual (IDM)
- ✔ International Framework for Dictionaries (IFD)
- ✔ BIM Collaboration Format (BCF)
- ✔ Model View Definition (MVD)

The previous section discusses IFC, and the following sections examine the other four in more detail.

The process standard — IDM

When you think about construction projects you've been involved in, try to recall how many various parties the client brought together. Was it clear what information each project team member required at what stage? How did the information manager communicate it? Did any problems occur because you didn't receive the right data or the right formats?

IDMs specify communication and data exchange requirements during a project's lifecycle. The IDM standard (ISO 29481–1:2010) sets out the method for capturing the process describing who, what, and when. Who needs to provide data, what do they need to provide, and at what points in the project?

The terminology standard — IFD

Have you worked on many international projects? Did language and terminology differences cause problems between offices or suppliers? The IFD standard (ISO 12006–3:2007) aims to enable models and systems used on projects to be language independent. In other words, it doesn't matter if a faucet is called a *faucet* in the United States, in the UK it's a *tap,* and in other languages it's a *Wasserhahn* or a *robinet,* everyone is talking about the same object.

So buildingSMART manages something called the Data Dictionary (bSDD) based on the IFD standard, as a reference library to allow the construction industry to share and compare consistent product information, irrelevant of its origin. It promotes the linkage and transfer of information in existing databases too.

The change coordination concept — BCF

How do you inform another team that something's wrong with the model? Two of the large BIM vendors, Tekla Corporation and Solibri Inc., proposed the idea of an additional open standard that would allow comments and snapshots in the model to improve communication across project teams using different platforms. BCF tries to resolve the problems that arise when merging and importing data from multiple sources.

The schema basically splits comments (from one team to another about issues in the model like clashes and design errors) from the model itself and allows another party to reopen just the references to precise locations in the model and not the entire BIM. buildingSMART agreed that this idea was worth adopting. At the time of writing this book, BCF is in the process of becoming a fully fledged buildingSMART specification.

The process translation concept — MVD

For a particular piece of work, how much of an entire information model are you going to require? In the same way as you used to generate just the drawings needed in 2D to explain the project, you can just select *model views* or portions of the BIM. In order to satisfy the varied information exchange requirements of the AEC industry, MVDs predefine the amount of the model required to support a specific stage of the project. buildingSMART calls these *subsets* of the model, and the MVD indicates what information is required.

The receiving party probably doesn't require the full BIM, so by appropriately classifying elements using IFC classification headings, party A sends only the relevant elements and information to party B. For example, a mapping of the FM basic handover MVD, which includes operational information, is a COBie spreadsheet (which we explain in more detail later in this chapter). Other MDVs for IFC 2x3 include the coordination view and the structural analysis view.

Reviewing the Key Standards

Standards are important to you because they set out the framework for a BIM process and the guiding principles that you should follow. In the following sections, we discuss more about the BS and PAS 1192 documents and their main highlights you should know.

Developing and updating standards

You may work with standards on a regular basis in your daily work, and sometimes you receive notification that an update to the standard is available. As the industry evolves and changes, the construction industry needs to review standards documents to reflect current practice and future direction.

To do these reviews, the standards organisations like ISO and BSI set up committees of industry experts. The B/555 Committee is responsible for standards within construction design, modeling, and data exchange. B/555 sets out within its roadmap the maturity sequence that the standards will go through from Level 0 to Level 3 BIM, in accordance with the Bew-Richards BIM maturity model (refer to Chapter 9 for more information).

Analyzing PAS 1192

In the UK, the PAS plays a fundamental role in supporting the objectives of Level 2 BIM. It not only specifies the requirements for achieving it, but it also sets out the framework for collaborative working on BIM-enabled projects and provides specific guidance for the information management requirements associated with projects delivered using BIM.

PAS 1192 parts 2 and 3 together with its older sister standard BS 1192:2007 and now younger brother BS 1192:4 are the only tried and tested standards that support the UK Construction BIM Strategy to achieve Level 2 compliance and the desired reduction in CAPEX outturn cost. In the following sections you get to know them personally and what part each plays within the BIM process.

Taking away from BS 1192

When the industry moved from paper and pens into CAD software, BS 1192 (first published in 1998) provided a guide for the structuring and exchange of CAD data.

The standard was revised in 2007 and given a new title: Collaborative Production of Architectural Engineering and Construction Information. The revision put more emphasis on collaboration so that the construction industry would effectively reuse data. It promoted the avoidance of wasteful activities such as

- Waiting and searching for information
- Overproducing information with no defined use
- Overprocessing information simply because technology allowed it
- Reducing defects caused by poor coordination and resulting in rework

PAS 1192-2

The need to understand future use of information and what happens farther along the supply chain led to the creation of PAS 1192-2:2013, "Specification for information management for the capital/delivery phase of construction projects using building information modelling." As the name implies, the document focuses on BIM in the delivery phase of projects.

A PAS document is a Publicly Available Specification, which is based on the principles of British Standards but proposed for adoption by an industry very quickly. A full British Standard would need all stakeholders to reach a consensus on its detail, but a PAS can get an entire sector of industry working to the same code of practice, rather than everyone trying to document their own processes and share them informally.

The benefit of using the PAS process is that a new industry standard that helps with a specific market need can be developed and released to the construction industry in a very short space of time. PAS 1192-2 was sponsored by the Construction Industry Council and is free to download.

BS 1192 is still at the core of PAS 1192-2:2013, and both supports and underpins the means of achieving Level 2 BIM compliance.

BIM: A standard framework and guide to BS 1192

To accompany the PAS standard, a helpful guide document covers the processes required for public project delivery, which the UK government has set as the initial target sector. It's one of a number of documents supporting the government's strategic objectives. These include

- ✔ BIM protocol
- ✔ Employer's information requirements (EIR)
- ✔ Construction Operations Building information exchange (COBie)
- ✔ A digital, cross-industry plan of work
- ✔ A digital, unified classification for the construction industry

For more information on the Level 2 BIM suite of supporting documents, flip to Chapter 8, where we explain these standards and documents in more detail.

PAS 1192-3

PAS 1192-3 is officially titled, "Specification for information management for the operational phase of construction projects using building information modelling." They sure don't make these names short!

Whereas PAS 1192-2 focuses on the delivery phase, Part 3 moves on to the operational phase; in other words, how the project team transfers data to the owners and operators of assets and buildings when they're in use. One of the most interesting aspects is how the asset management team then transfers information from the design and construction model to an existing enterprise system.

Two terms we mention we introduced in Chapter 1: we use project information model (PIM) and asset information model (AIM) here to separate the information model in the delivery phase from the operations phase. AIMs can also represent an existing building or project.

We can describe PAS 1192-3 as a partner document to PAS 1192-2. Part 2 has a clear sequence, whereas Part 3 describes events in the lifecycle of an asset that could happen in any order between the point of handover and eventual discard or demolition of the asset.

To use PAS 1192-3, you transfer information to lots of different stakeholders in the built environment. PAS 1192-3 focuses on the operations and maintenance of assets, so it's intended for asset managers and owners. It makes a clear distinction between asset management and facilities management and the terms used. PAS 1192-3 uses the term *asset* to refer to concrete physical resources.

Some of the information received for the AIM is likely to come directly from computer-aided facilities management (CAFM) systems, including tracking of repair items and the causes of asset faults.

 If you're involved in asset management or interested in the operations phase of assets and buildings, then you'll want to check out ISO 55000 (replacing the former PAS 55 documents). This series of international standards provides the international standard for whole lifecycle physical asset management.

BS 1192-4

The Construction Industry Council (CIC), on behalf of the UK BIM Task Group, sponsored the development of BS 1192-4, making it available as a free download. BS 1192-4 is a code of practice rather than a specification standard. A *specification* refers to absolute requirements that must be followed to achieve a specific outcome, via actions that are considered to be aligned with current accepted good practice. A *code of practice* means that the document recommends good practice and guidance with a degree of flexibility that is less rigorous than a specification but also offers reliable benchmarks.

BS 1192-4 replaces another schema called COBie-UK-2012 and is intended to assist the procurement and maintenance of assets, making it relevant for portfolio and facility managers to specify their expectations. For the design team and others producing information, it helps the preparation of concise, unambiguous, and accessible information. It defines the methodology for the transfer between parties of structured information relating to facilities, including buildings and infrastructure, and draws on experience gained on a number of UK government early-adoption BIM projects, such as Cookham Wood Prison.

You can find more information about COBie in the next section.

Coordinating Information

Standardized information is at the heart of BIM and especially at the heart of the UK BIM strategy. The suite of documents that defines Level 2 BIM requirements is built around that principle. Regardless of what software you are using or level of BIM maturity, a simple mechanism for data transfer is required. That mechanism is COBie, which acts as a bus to take information from one place to another. The next sections look at COBie in further detail, including what it is and what it does.

Providing the framework for data exchange

Originally developed by Bill East of the United States Army Corps of Engineers, *COBie* focuses on information rather than geometry and is described by the UK BIM Task Group as being "simple enough to be transmitted using a spreadsheet."

COBie allows the project team to organize, document, and share information about assets in a standardized way. Simply, COBie allows the project team to capture the important data about a project in a really clear format, so that facility management and owner-operators can reuse it.

The spreadsheet format allows for high interoperability, and your COBie output could contain product manufacturer data, warranties, and equipment schedules.

Working with COBie

The purpose of COBie is to capture critical information for owners and operators to assist with the management of their assets. BS 1192-4 states that the "process of exchanging COBie deliverable should be integral to the whole facility lifecycle to maximise the benefit and efficiency of the employer-side pull for information."

With the UK's 2016 Level 2 BIM mandated deadline date fast approaching, the government has promoted COBie as the recommended open standard data format.

 COBie is a subset of the BS ISO 16739 IFC, documented as a buildingSMART MVD, which includes operational information. The 2014 version of BS 1192-4 documents best practice recommendations for the implementation of COBie. (To use its full title, "Collaborative production of information Part 4: Fulfilling employer's information exchange requirements using COBie — Code of practice.")

In association with the open BIM network, NBS tested whether the buildingSMART IFC file format was capable of supporting the creation of COBie datasets. It did this by running a trial with the help of a number of principal UK contractors. The resulting IFC/COBie Report 2012 is available to download from www.thenbs.com.

Chapter 12

Encouraging BIM in Your Office or On-Site

. . . .

In This Chapter

▶ Transforming traditional workplaces into digital-centric ones

▶ Putting together the perfect BIM team

▶ Using training to bring everyone on the BIM journey with you

. . . .

*B*IM is a process made up of many steps, and it needs the right platforms and software to support that process. But BIM isn't going to work if you don't also involve the right people. Whatever your role in the construction industry, if you're a manager, client, designer, contractor, or building owner, you need to ensure you have a team around you with the skills to make BIM successful and meet your objectives.

Another way to look at BIM is to think of the following equation. BIM is a combination of your project's lifecycle (building), its embedded data (information), and the geometry (modeling). To support all that, you equally need a blend of the workflow you'll use (process), the project team (people), and the software to apply everything (platform):

$$\text{Building} + \text{Information} + \text{Modeling} = \text{Process} + \text{People} + \text{Platform}$$

Your objective may be cost-cutting carbon savings, operational handover, or greater practice efficiencies, and you need to be confident that everyone is working toward the same goal using BIM to realize it. Throughout the book, we explain that BIM isn't about technology; it's about a new set of soft skills. Each member of the team needs these skills to make implementation successful. This chapter is all about showing you how to achieve that.

Adopting BIM as a People-led Process

You may have encountered quite a lot of mixed messages about BIM adoption and implementation. Perhaps you've heard stories and anecdotes about successful BIM on big projects and in small offices, but cutting through all the noise can be difficult. Finding out the information relevant to your situation isn't easy because everyone's story is different. But the one common factor is that people always rally together to get the job done.

We hope that you're thinking of BIM as a process, but don't just think of it like a production line made up of computers, factory robots, and new hardware and software. BIM is about a series of improvements in the way people work and the way you work with other people. For you, this may be as part of a team based in an office or on a construction site. BIM is the methodology for driving better working practices and improving communication across the entire industry. In other words, BIM is a people-led process.

This chapter pinpoints the key soft skills that we think BIM teams need. If you're looking for help with team management or team building, we suggest that you check out the newest editions of *Project Management For Dummies* by Stanley E. Portny, *Managing Teams For Dummies* by Marty Brounstein, and *Psychometric Tests For Dummies* by Liam Healy (all published by John Wiley & Sons, Inc.).

It's also vital that all team members understand what they need to do and the definitions of their roles and responsibilities. Check out Chapter 17, where we cover lots of BIM-related jobs, for more information.

Making the most of soft skills

The increase in BIM projects and implementation ahead of government mandates or commercial drivers means a BIM skills shortage in a number of countries. The skills shortage involves some technical knowledge and core technology skills, from detailing built connections to embedding models with property data. You may have heard politicians describe them as *hard skills,* things that are academic or reflected in qualifications, because you can teach and develop these kind of aspects on the job.

It's far harder to teach the other half of the skills problem, the soft skills. *Soft skills* are abilities like

- Problem-solving
- Initiative
- Creativity
- Time management

These soft skills are the core talents that determine the behaviors, atmosphere, and success of groups in an office or on-site environment. Another way to describe them may be *work ethic*.

What is it that makes you tick? Why do you work the way you do? We go into these questions in a lot more detail later in the chapter. Think in this way:

- ✔ People interacting with technology: cool.
- ✔ People interacting with *people,* supported by technology: really cool!

Leading BIM implementation

One of the key skills necessary for successful BIM processes is leadership. BIM is always a collaborative effort, but especially in the early days of implementing new procedures and protocols, a clear and decisive leader is invaluable. If your role is leadership, you need be a point of contact, a role model to lead by example, and a motivational figurehead when challenges occur. Here are some essential leader's soft skills:

- ✔ **Effective communication:** Making your points heard and understood clearly is critical. Sometimes you have to communicate difficult or sensitive things in a calm manner, or be firm with people without getting angry or flustered. More than anything, you need to be honest and clear. That combination can make you a very persuasive presenter.

  People don't have time to read long reports any more. Communicating something like a business case is no longer about 400-page papers or death by slideshow. You need to produce key summaries, short visual representations, and clear infographics that get to the point and provide the facts.

- ✔ **Listening:** Hand-in-hand with communication is the ability to listen. If you're a leader or key advocate for BIM then people will want to talk to you, and you need to understand their point and respond to it. One great technique is to clearly repeat back what you've heard in order to confirm your understanding.

- ✔ **Management:** More than likely as a leader you have management responsibility — perhaps for BIM teams or for people with their own line reports. Only by spending regular time with your staff can you ensure that BIM implementation is going well on the ground. Also, by coaching those reports and developing their skills, you can embed your vision for BIM.

✔ **Availability:** This skill can be a challenge. We know that you're probably busy, and if you're in a leadership role, you'll know all about being busy. However, being available and active to people who need your help, opinion, or advice to move the game on is vital. If people see that you're committed to BIM and dependable in a tough situation, you can instill those same traits in them. Basically, lead by example. The best examples are being first to arrive on-site in the morning or the last to leave the office at night.

Although you want to develop hard skills like IT and technical construction ability, because hard skills are essential, successful implementation of BIM relies on better soft skills across all sectors of the industry.

Recently, McDonald's conducted some research in the UK that found that soft skills contribute £88 billion to the UK economy, and with an increased reliance on service industries the value they add is set to increase to more than £120 billion.

The research also found that without more investment in soft skills, in 2020 more than half a million workers will be held back by a lack of them. Education can often put a focus on you achieving a specific qualification or getting the best exam results. Employers globally are demonstrating that you really need a combination of hard and soft skills. The classic wisdom is that *exam results get you the interview, but soft skills get you the job.*

Aiming for the ideal team

Many great methods and approaches define the perfect team. If you've been in business or office environments much, then you've more than likely heard a lot of theory about team building and personality types. The business world and manufacturing industry have learned quickly that a combination of diverse and complementary skills make for the best outcomes. However, if you have any experience of working in teams, you know that a mix of opinions and characters can also lead to conflict and clashes.

In terms of the ideal BIM team, here are some key features to look for:

✔ **Teamwork:** Pretty obvious, right? A good team needs teamwork skills. No kidding. . . . In reality, the ability to fairly and cooperatively work in a team is a real soft skill. Certain people are natural team players; others are leaders and some like to work alone. In BIM, everyone needs to pull together to achieve a common goal and form a true team.

✔ **Interrogative and critical skills:** BIM produces a lot of data, and sometimes all you need is the facts. How do you ensure you're getting the relevant information and asking the right questions of the model? Having a team that has the skills to contribute key facts and to analyze and critique the data without bias is a recipe for quality information outputs.

✔ **Risk management:** Understanding risk is such a core skill in BIM. From identifying safety issues to consider in the design phase to avoiding construction risk or digital security management with sensitive project data, you need a team that can mitigate risk and consider alternative approaches.

✔ **Cooperation:** This skill is probably the softest skill of all, but no less important. Do your people get along with other people? Do they get along with each other? You need to make sure that your team members have attitudes that foster great relationships. A lot of great ways exist to break the ice between people, and no one knows your team better than you to work out whether a game of football, a fun evening out, or a boardroom meeting to thrash out issues is going to have the biggest impact.

How big should your team be? That's one of those impossible questions because the answer depends on so many other factors like project scale and value. One of our favorite stories about the search for the ideal team size is about Jeff Bezos, the founder and CEO of Amazon. He has a "two pizza rule"; teams should always be small enough that two pizzas would be enough to feed everyone. He suggests that having more people in the room actually has a negative effect on communication. These teams may be part of bigger departments or management structures, but they're small enough to get things done quickly.

You may be fortunate enough to have all the skills you need in-house. For most people, to build the ideal BIM team you need to think about recruitment. You may be at a managerial level or involved in HR processes and have a direct influence on this. If not, be brave. Talk to your managers or HR directors about what skills (hard and soft) you think the business needs to move forward with BIM.

Encouraging BIM Champions

At some point in the past, the three of us have all been fortunate enough to have someone explain BIM to us and spark our interest. Thankfully, that conversation was early enough for us to have spent the last decade really investigating digital construction and how BIM should work and how real projects can demonstrate its benefits. Over that period we've developed the knowledge and skills to become not just BIM advocates but true champions of it. Some people even have the job title "BIM evangelist" because they're spreading the good news of BIM.

These sections recognize the soft skills that make for great BIM uptake across teams and businesses and the ways to encourage the people who work with you or for you to be BIM champions. Here are three skills of a BIM champion:

- ✓ **Enthusiasm:** This skill refers to a "can-do" attitude. The champion considers nothing impossible and can put together a plan to achieve your goals. The champion is not only excited about BIM and its potential in the future, but understands how it changes things for your business today.

- ✓ **Adaptability:** BIM is a quick-moving target with new technology and documentation on the horizon all the time. The term *jack-of-all-trades* used to have a negative connotation, because it implied master of none. In today's construction industry, this isn't the case. Someone who can see the whole BIM picture and master skills across all levels, from object modeling to project management, will thrive.

- ✓ **Determination:** BIM isn't without its challenges. A BIM champion who can overcome challenges and obstacles that you may never have encountered previously requires resourcefulness and hard work. Working out how to fix things when they go wrong and continue pushing on is especially important.

Sometimes the best soft skills come from your personal life experience, such as resolving a family conflict, teamwork on college sports teams, or managing your time around childcare. What lessons can you take from your daily life outside of work?

The other benefit of soft skills is that they're *transferable*. No matter how your career develops or what role you have in the industry, soft skills are incredibly useful and you can always apply them to new kinds of work.

The skills that make a great BIM champion are often down to experience and the right personality type. The following sections are about what to look for in your colleagues to pick out future BIM champions.

Reflecting on the right experience

Say that you've been given a task that you don't know how to approach, and you need some good advice about how to break it down. We're going to hazard a guess that you can quickly think of someone you can ask about the task; for example, a mentor or manager you really look up to and regularly gain insight from. What sets that person apart? Your BIM champions will be able to use their experience and apply it to day-to-day BIM implementation:

✔ **Problem solving:** Every day will bring challenges, and you need team members who can solve these problems. BIM champions should be able to apply logical thinking to a problem and make contributions to your BIM teams that lead to practical solutions by taking the lessons learned from past projects.

✔ **Autonomy:** This just means working independently. Your BIM champions should be able to take full responsibility for their work, including any mistakes. BIM champions need to have confidence in their growing ability and BIM methods to achieve the results required without asking questions all the time.

✔ **Lifelong learning:** BIM champions not only need to learn from previous experience, but they also keep researching, reading, and attending events to increase their knowledge, so that they stay passionate about personal growth. You may also hear people call this *continuing professional development (CPD)*. This is something you should keep doing too.

✔ **Prioritization and planning:** Something that only comes with experience is realizing how much time some tasks take. Being overly ambitious with deadlines and overpromising things is easy. Handling lots of projects or tasks at the same time and working out the priorities is a real skill, and one that BIM champions should demonstrate.

BIM advocates can come from any discipline in the construction industry, but don't forget about the transferable skills people bring from alternative sectors. Some join the construction industry with varied backgrounds like IT, accounting, or academia, but you still can find great team workers, trouble-shooters, or project managers in those environments.

Watching for the right personality types

Say that you have an urgent task and you simply can't achieve it in time. You need to delegate it because you have equally urgent but more important priorities. Quickly, think of team members you work with whom you can approach with that request. Perhaps you approach someone who just gets things done or is incredibly good at organizing her time. Here are some additional BIM champion skills that are about people's personalities:

✔ **Self-motivation:** If you're in a managerial position, how do you feel about people taking the initiative to make decisions and move things on? Workers who are self-motivated and passionate can make your life much easier. Of course, a balance exists, but BIM champions should know when to ask and when to make progress with their work.

✔ **Time management:** Of all the soft skills we discuss in this chapter, people who can manage their time are perfect for BIM processes. Construction has always required submission delivery under pressured deadlines in the middle of the night, so composure in the face of daunting timescales is ideal.

✔ **Productivity:** BIM champions need to balance time management with productivity. This just means how much gets done because of people's effort. How quickly can you get that model completed? What is the client expecting in the presentation; can you pull that together today? The perfect BIM team needs lots of people who can produce great work on demand.

✔ **Respect:** You'd be amazed by how many rude people are in business and industry. You probably have stories of being open-mouthed by things people have done and said. A fundamental rule is that everyone is polite with co-workers, especially clients. Respect is something all BIM champions should have.

Equally, you can probably think of someone you'd never approach with these kinds of tasks. Why is that? Is she disorganized? Disrespectful? Dismissive? Disruptive?

Changing Hearts and Minds

You should feel confident that anyone can become a BIM advocate and support your implementation of new processes, no matter what position she starts from. By applying some simple methods, you can encourage all members of your team, however negatively they may start out.

Everyone in your team will be at a different stage of BIM understanding and enthusiasm, and one of the biggest challenges in BIM is working with difficult people. In these sections, we provide some tips to help you bring teams together and realize that everyone has something to offer.

Making the early adopters effective

When you first heard about BIM, perhaps you just got it straight away. Many other people have done the same thing, which is great, but watching the slow rate of change in your offices, sites, and the wider industry can be frustrating.

"Why can't we be more like [insert more advanced industry here — aeronautics, technology manufacturing, automotive design]?" Being ahead of the curve like this can actually have a negative effect on people's attitudes, sometimes making them hasty or reckless in changing things or, worst of all, making them think of leaving your business to find BIM opportunities elsewhere.

If you have team members who are keen to race ahead, how can you make the most of that? Perhaps this is a good way to describe you too. The key thing here is to make these individuals feel valued and that they're having a direct impact on process and procedure. For example, you could

- ✔ Give them responsibility for a discrete aspect of the implementation.
- ✔ Let them document the processes in their own way.
- ✔ Ask them to communicate the benefits to other staff.

Probably the best advice is to use the early advocates to promote and embed the new processes and to encourage everyone else, rather than having to do it all yourself.

Pushing the late bloomers

You encounter team members who take a bit of time to get their head around BIM and your plans to change existing processes. Perhaps you know someone who still thinks BIM is just 3D CAD and doesn't really understand the data and embedded properties. Maybe some people recognize what BIM is trying to do as a process, but they just like using paper drawings or using their own file-naming system.

We're guilty of this too. The digital realm sometimes doesn't feel as if it has fully solved the way people interact with physical materials quite yet. Maybe you're reading this as an e-book or online, but we know that many people still prefer a paper copy.

The following methods can help you to encourage adoption of BIM concepts in the late bloomers:

- ✔ **Understand the impact.** Sometimes grasping the effect on the team can mean realizing that because of an old process a co-worker has extra work to do. Think carefully about the people you're trying to help and what's important to them. For example, if they're keen on sustainability, point out the environmental benefits of BIM.

✔ **Connect people.** Nothing beats a friendly colleague helping you with something you don't understand. When instructions and mandates come from management or leadership, they can seem daunting — or, worse, you can feel that the manager hasn't considered the day-to-day realities of work. Having a peer-level co-worker as a BIM champion can really improve acceptance.

✔ **Plug the gaps.** Often, people stick with processes they hate because doing so is easier than learning something new or having to start again. Offering a training course instead of asking people to self-teach can make a big impact. We discuss more on training in the later "Training and Supporting Everyone" section in this chapter.

Motivating the cynics

You probably know some people who think BIM is just a load of hype and they've heard it all before. How can you motivate the people who just don't seem to want to hear about BIM? Here are some skills you can apply to help you in this situation:

✔ **Make it real.** You can demonstrate the cost implications of not implementing BIM. As harsh as it sounds, business is most often about profit, and the bottom line and people's jobs are part of that equation. Showing that BIM improves business processes and design/construction outputs can reveal what it takes to make a more stable or profitable business, and that can result in increased job security.

✔ **Explain the benefits.** You can start to discuss the simple benefits of BIM, which we discuss in Chapter 13. However, explaining isn't the same as displaying them on the ground. For example, if you can show how the properties in BIM objects make information retrieval easier, you can compare the speed of that process to finding data in vendor catalogues or stacks of paper drawings.

✔ **Respect experience and tradition.** Sometimes, change is the scariest thing. People can easily let their minds run away and think that because of BIM their skills are irrelevant or that they simply can't learn the new technology quickly enough. This is especially true for older staff. Demonstrate your respect for their experience and involve them in critiquing and informing your BIM strategies. See which traditions you can adapt, rather than remove. Change roles to make people educators, not dinosaurs.

You never want to think about it, but despite your best efforts, sometimes some people won't fit into your new collaborative BIM culture. They may not want to join your team environment or they may lack the respect of clients, co-workers, or customers. Don't let one person's disruption derail the success of your BIM strategy. We're firm believers that everyone has a role somewhere, but you also need to know when to draw the line.

Jumping the barriers

What are the obstacles and barriers that you need to overcome to implement BIM? A few soft skills are fundamental to BIM, such as the ability to share and collaborate, and your team needs to have that mind-set. You've probably encountered processes that amaze you because they're inefficient or outdated, but they're "the way we've always done things." This section helps you to overcome those hurdles and assist people to change the way they work. We also show you how the *way* that people work is transforming.

More than anything, BIM moves a lot of detailed work to the front-end of the project. Instead of a design team handing over its information to contractors at tender and the contractors continuing to develop design and construction strategies, the entire project team needs to work together from the early stages to ensure that they produce and collate as much information as possible to solve problems in the digital world before encountering them in the real world.

Sadly, moving more work to the front-end can result in conflict and confusion if you don't manage it correctly. We cover a lot of the legal and commercial issues BIM creates in Chapter 14, but in terms of day-to-day management you may need to use and develop your conflict-resolution and negotiation skills. Most of the time, people want fair treatment.

Here's a great story, attributed to Mary Parker Follett, about resolving conflict that shows sometimes a 50/50 split isn't what people really need. A mother finds her two children fighting over a single orange. To resolve the argument, the mother cuts the orange in half and sends the children away. Later, she finds them both still unhappy. It turns out that one child wanted all the peel to flavor a cake and the other wanted to squeeze the pulp to make orange juice. The mother's solution wasn't a suitable one, but both children could have had what they wanted if they'd explained their needs.

Transforming the workplace

BIM can be transformative in the built environment industries. We can't really emphasize the power of BIM enough. It's not just hype and concept. By increasing access to information at the point where project teams most need it, you can make major changes to the way people work:

- Flexible working hours in global collaboration
- Remote working online, in the cloud and via virtual private networks
- Trust and confidence in shared data and its security

As an industry, construction engineering is still uncomfortably imbalanced in terms of gender and minority representation, and getting rid of discrimination needs to be everyone's priority. You can make the construction industry more diverse and equal than it's ever been! A successful BIM team needs to be diverse.

The built environment isn't unappealing to any group; you know well that it's full of varied and different roles that need skills in art, math, communication, science, engineering, creativity, technology, craft, manufacturing, and design. But still something hinders everyone feeling like they have an equal opportunity to succeed. If you're in management or recruitment, you have the power to change an industry, one new hire at a time. If you're not at that level, you can just make sites and offices friendlier places to be for an individual, whatever that means for someone. You can make construction the world's model for equality and diversity, not an embarrassing relic of the past.

Training and Supporting Everyone

You may have spotted that hundreds of providers offer BIM training at all levels, from beginner sessions to full certification and college degree courses. Do you need to join one of these courses? Do you need a BIM degree to be a good BIM manager?

Everyone's different. You or one of your colleagues may really benefit from being able to analyze all the principal BIM documents or critique existing case studies in an academic environment.

The danger is that making BIM too academic is easy. If you want to write your PhD on "An evaluation of team dynamics during BIM implementation," we're not going to stop you. What we're here to do is help you develop a mostly practical skill: being able to collaborate with many other people to successfully model the built environment and pack it full of data you can retrieve when you need it most.

After your colleagues are on the road toward BIM implementation, you need to train them on the platforms that you'll be using, support them in newly created roles, and maintain best-practice processes for many years to come. The following sections take training, new roles, and new activities in turn and show you how you can build the initial excitement of BIM into real, long-term change.

Training at every level

From the ground up, every single person involved in your BIM processes needs some form of training. Some of this training is in the hard skills and technical knowledge, such as mastering a new version of a software platform or understanding the core BIM protocols and your particular government's documentation on construction strategy. Sometimes you can deliver internal training that meets the needs by drawing on internal expertise. Other times you need to look to external providers.

A critical need is to get people up to speed quickly. That could be as a result of new software or hardware implementation, a new process, or perhaps because the staff member is new to your existing BIM systems. Always make training specific to your operation and ask providers how they can adapt their offerings to tailor content to your business. If you just receive basic and generic training, you may not be getting to the heart of how a particular platform or protocol can generate new efficiencies in your projects.

You don't *need* a qualification to become a BIM user, a BIM advocate, or even an expert in BIM. Nothing beats real-life experience to understand what a change BIM makes and how to generate these new efficiencies. We can sum up this life experience with the phrase "human teams, paper reams, rainy sites, and late nights."

Emerging into new job roles

One of the major changes that results from BIM is the development of new and unfamiliar responsibilities, many of which we cover individually in Chapter 17. Some familiar themes run across all of the roles and responsibilities, however. Your team needs these skills to take on roles like BIM management, BIM coordination, and data analysis:

✔ **Sharing data:** Your BIM teams may need to move from a traditional responsibility for a discrete piece of work, to a collaborative and open environment where issues are discussed publicly in the context of other people's work. This is integral to resolving errors and clashes between different parties on the project team as early as possible. The environment should never be embarrassing, shaming, or scary. Everyone should feel confident to input, speak up, and contribute.

✔ **Staying up to date with industry news:** BIM teams need to be informed about where the industry is heading and what impact technology innovation or legal changes can have on their work. Because many BIM projects span multiple years, keeping current with a changing industry is vital. Your teams should look to reliable newsfeeds, social media, blogs, and industry publications for the latest updates.

✔ **Being self-motivated and keeping a positive outlook:** Whatever roles your project team members have, they should all be self-driven and prepared to work hard. BIM isn't a magic bullet that makes built environment projects easy. Just like they have for hundreds of years, construction projects can get tough. Your team must be able to maintain positivity, no matter what state the project is in, or what the weather is like on-site. Everything will be worthwhile when the project is successfully delivered.

Maintaining new activities

BIM implementation can take a long time. As technology and documentation improves, making BIM a success in your business is easier, because there's more support out there for you — and it's going to continue getting easier over time. But don't underestimate that some processes and people will take longer to embrace change.

Then, after you have BIM processes, platforms, and roles that you're happy with, you need to maintain the activities and work flows that you've put in place. You want to keep morale high among your colleagues. Think of how differently your team might feel in their first BIM meeting compared to when they're in the depths of a project six months later:

Day 1: BIM is going to change everything! Yes! Let's do this!

Day 200: Is BIM changing everything yet? Is everyone still up for this! No?

Wait, what happened? Those initial meetings may have been great and everyone was keen to understand their role in the process. However, after a year of trying to change existing procedures, document protocols, and rename thousands of files or BIM objects, your early enthusiasm can start to fade. This is totally normal.

Staying committed to BIM is important. Think of it like exercise. In the same way that you need to commit to an exercise and nutrition regimen, even during the tough bits of the diet or the hardest workout, you can appreciate that the benefits far outweigh the temporary pain. Consider that your business or project probably needs to lose a bit of weight it's carried around for a while that doesn't restrict it completely but makes it sluggish and slow. When things get busy and the business needs to run at peak capacity, it gets tired quickly. BIM isn't a quick fix; it's a lifestyle.

What is the exercise regimen for your processes? Just like human health, when your business is super-fit, you reap the benefits in the long term.

The more creativity you have around you, the more likely you are to keep finding new ways to do things better. Try to surround yourself with innovative thinkers who like brainstorming and can apply those ideas in practice. Experience tells us that creative people have a direct link to effecting change in the workplace and implanting new processes.

You can always find someone to help you. In Chapter 22, we show you lots of great resources to support you in your BIM implementation and delivery. In particular, a great community of BIM professionals, advocates, and users exists online. You're not alone in the experiences you encounter while working with BIM. We can't emphasize this enough.

Chapter 13

Developing BIM Plans and Strategies

So you've won over some key decision makers and they've asked you to make a case for BIM, but just what does having a BIM strategy really mean? How do you justify investment in BIM technologies, software, hardware, new room layouts, or even new staff? What's the real benefit to the company and how quickly can you demonstrate return on investment (ROI)?

When the UK government looked to adopt BIM, it had to demonstrate how BIM would result in significant cost savings, increased project value, and improvement in carbon performance, all through more efficient access to shared information about built assets. Perhaps you're a BIM manager or BIM coordinator and it's now down to you to make the case for BIM and set out your strategy. In this chapter, we explore where to start and what you need to consider to get going.

Getting Started with Your BIM Strategy

Before you start, ask yourself: What is your company trying to achieve with BIM? Perhaps you want to produce better coordinated designs and assist with budgeting and pricing. Maybe you want to improve decision-making processes through better visualization of the end deliverable and what-if scenarios. Do you want to move away from paper-based handover documentation and produce digital health and safety files and operation and maintenance files? Maybe you want all the above. Unfortunately, no one-size-fits-all, off-the-shelf approach exists. Each organization and project is different, and therefore your digital journey and strategy is unique to you.

In the next sections we examine some important strategies, including what to pre-plan, how to embed BIM into your everyday work, and how not to focus too much on technology.

Undergoing some pre-planning

Know your benchmark and your starting point in terms of both capabilities and which key performance indicators (KPIs) you want to measure success against. Here are a few areas to review and consider before you can figure out which strategy is right for you and your organization:

- ✔ **Get top-down support from senior management and key decision makers and communicate it.** Although you need the enthusiasm from the bottom up, you require the support from the top to make sure that the message carries weight through the organization. Demonstrate that improved access to shared, accurate information will reduce costs in the long term, but be realistic in your promises.

- ✔ **Consider cyber security for your valuable digital data.** Today low-life thieves make off with more than just physical possessions. They also focus on digital data. In the wrong hands, crooks can make money off your sensitive information and intellectual property.

- ✔ **Determine the current archiving and document control protocols and procedures.** Getting rid of those space-invading file cabinets not only frees up office space but also makes you think about how and where you can store your digital data.

- ✔ **Make sure that your staff members receive the required training for the job.** Figure out whether you need to use an outside resource or if you can deliver the required training in-house. Do remember that everyone is human at the end of the day and different people learn at different rates, and some may require a bit more help than others.

- ✔ **Identify who'll maintain your strategy and where it will live.** Will the BIM manager be in consultation with the rest of the team? Remember that your strategy should be a live document that will be updated, and will require an owner who will carry out this task.

- ✔ **For IT, get support and buy-in from the IT team or manager.** You may be faced with software and hardware upgrades, so having the people who carry out these tasks on your side is always a wise move.

If any of these areas aren't up to scratch, make sure that you bring them to the table early and make use of your internal BIM champions to support you in your endeavors. Refer to Chapter 12 for help in identifying your BIM champions.

Embedding BIM into everyday work

To go from having no BIM strategy to it suddenly becoming part of everyone's jobs, you need to manage everyday expectations. Your first BIM project is unlikely to be transformational; getting it optimized and embedded will take trials and refinement. So remember that implementing BIM is a marathon not a sprint.

For now, don't worry too much about technology, the cost barriers to implementation, or the business benefits that BIM brings. For now, focus on BIM being a process and behavioral change program.

Barack Obama spoke about change in his election campaign. Winston Churchill said: "To improve is to change; to be perfect is to change often." Both Sir Michael Latham and Sir John Egan, in their respective reports, made recommendations to change construction industry practice. However, with human nature and social behavior being what they are, people often stick to what they know best. That's all very well, but would humanity have put man on the moon if people hadn't pushed the boundaries? Can you imagine life without the Internet? The fact remains that to really innovate, people need to adapt and change. Change is inevitable, and unless you're willing to be open to change, things can become stagnant.

Embed BIM into your current day-to-day workflows, not as a separate entity. Here are a few tips for embedding BIM practice into everyday life:

✔ Ensure that senior management take responsibility, because a change in working practice always starts at the top.

✔ Establish and communicate a common understanding of what BIM means to your organization and customers. Give it real purpose and focus on the *why?* rather than the *what?*.

✔ Identify key internal BIM champions.

✔ Start early.

✔ Make sure that technology, software, and processes are interoperable and integrated as a whole.

Understand that embedding BIM into everyday work life requires both time and patience. In the next sections, you consider when the change transition needs to take place and where to start on your journey.

Making the change

BIM is more than just software; rather, it's a combination of technology, processes, and people. The latter component is a bit of a paradox. For example, people can be an organization's biggest asset but create the greatest barrier to BIM implementation. An organization provides a wealth of collective knowledge. With knowledge comes power. But knowledge also brings preconceived ideas about how you should do things. After all, you've always done it that way, right?

We aren't advocating change for change's sake. The wheel is round and has been that way for a long time. It's been tried, tested, and acknowledged as the best configuration over many generations. However, the industry is moving into a new digital age, and therefore you need to reevaluate how you or your organization currently does things. Change doesn't have to be instant, it doesn't even have to happen overnight, but any change will benefit from a timely start.

Hasty implementations often lead to the wheels coming off. Make sure that you acknowledge that change brings uncertainty and resistance. People can feel threatened by any change, whether culturally or within processes. This resistance is often due to a lack of communication, trust, or even understanding. So early on, initiate the dialogue as to why the change is happening.

Make the explanation inclusive, and use business language and not technical jargon. Keep your words simple, easy to understand, and familiar, because already BIM is turning into a specialist subject.

Cultural change

When the industry changed from the manual drawing board to CAD systems, many a designer threw pens out of the cart in anger. On this occasion with BIM, you hear the sound of frustrated designers slamming the computer mouse down in anger, as the construction industry goes through a similar learning curve and change process, but more fundamental is the cultural change. Perhaps, just to add insult to injury, a lot of change is happening at once, when you consider new plans of work, new classification systems, and a move toward cloud-based storage solutions.

Make sure you include open communication

BIM brings about fear of the unknown as an organization develops the necessary adjustment to meet objectives. You can hear the gnashing of teeth and the plaintive cries: "What if I'm not up to it?", "This is a waste of time!", "What impact will this have?", and "What's all this costing?" These are all legitimate feelings and concerns, and recognizing the training that staff may require in order to gain confidence is important. Some companies are leading the way at the cutting edge of BIM implementation. Others, however, are at the start of that journey and moving in the right direction.

A BIM execution plan (BEP) document is a key way to communicate to the rest of the team how you'll be implementing BIM. For those individuals perhaps unfamiliar with the new processes that BIM brings, the plan provides reassurance and understanding. Refer to the later section "Combining Information into a BIM Execution Plan" for more information about this plan.

Rather than just telling the team about the change that you envisage, actually take the time to explain the decision-making process and why change is needed. We suggest that you create

- ✔ A simple companywide BIM 101 awareness communication (road shows or webinars are even better — keep them short and jargon free like a TED video)

- ✔ Some boilerplate BIM FAQ, addressing "What is it?" and "How will it affect my role?"

- ✔ A network of BIM change agents who'll cascade the message

Forgetting about the tools

Although technology plays an important part, at this point in your BIM project don't get too hung up on the particular software vendors or software application. People get caught up on the tools and not necessarily the processes or what the tools are actually generating. Take a step back and evaluate the process and what deliverables these tools can help you with.

Software is just a tool that requires:

Inputs + A process or use = Output

For example, think about a specification tool. Architects and engineers embed and capture company knowledge and expertise from previous projects, which is the input. Architects and engineers then use the tool for the process to write specifications and link the project specification to the object-based project models for coordination/additional parameters. The output then is an outline specification at one stage and a full specification at another, then perhaps an as-built specification at the end of the project.

Setting Up a BIM Business Case

As you prepare to implement a BIM project, you want to understand your drivers for BIM. *Drivers* are simply your ambitions for BIM and what you want to achieve from it. Do you want to implement BIM to win work or to increase efficiency? Ask yourself what the BIM project will mean in terms of money savings — reducing risk and adding value.

Your *business case* should be a well-structured document that answers these types of questions and captures the reasoning behind implementing BIM on a project. Think of it as the argument or a convincing case that you can put forward to senior management or decision makers that have the authority to approve or action it. Don't think purely in isolation. How will the business case fit in as part of a wider company change program, such as integrated delivery?

Business drivers include the following:

- ✔ **Return on investment (ROI):** Increased efficiency, reduced alterations and change orders, accurate costs, and improved timescales. Keep it simple and relate it back to your current business objectives. Your BIM strategy should be in concert with these and not a new layer.

- ✔ **Implementation time:** Think about a series of horizon lines such as awareness, mobilization, trial projects, and embedding.

- ✔ **Objectives:** Have a clear statement of objectives and outcomes.

- ✔ **Purpose:** Whose business case? Clients'? Contractors'? Users'?

- ✔ **Affordability:** Can you afford to do it? Can you afford not to do it? What's your competitor's intent?

- ✔ **Salaries:** How will BIM affect salaries?

- ✔ **Training:** How much investment is required?

With the business case in hand, turn your attention to the tools that can help you implement it. In the next sections, you consider the key points when investigating which BIM software and platform is right, and after you choose one, know how you go about implementing it in your organization.

Choosing BIM software and platforms

When making a decision about which software and platforms to use, seek advice not only from the project team but also farther afield. As well as talking to software and technology providers, have a chat with other companies and practices to see what they're using. Webinars, social media, conferences, and trade shows are all good avenues to do your research.

Retooling should be a result of having reviewed your business processes. Align retooling with your business needs, not the opposite way around. When considering which option is right for you, consider the following:

- ✔ **Solution:** Despite what the vendors may tell you, a successful project requires a number of different solutions to deliver the end goals because no single package can cover everything.

- ✔ **Investment:** Hardware, infrastructure, software, training, and related staff non-utilization can be costly. Remember when calculating your return on investment that the payback will be over a number of projects, so don't contribute all the costs to your first BIM project. Also remember that some free tools exist.

- ✔ **Version:** Nothing lasts forever, and sadly this is true of supported software. Whatever version you use, ensure that the whole team is using the same or that it provides you with the interoperability you require. Think about using open BIM file formats, which help you in the long term; for example, Industry Foundation Classes (IFC).

- ✔ **Management:** Consider who'll be responsible, and how, for the management and implementation of BIM technology, including deployment and software license management.

Implementing the software and technology

To determine the processes that will be required for a successful, painless, and stress-free implementation of software and technology, get together with your team and determine your requirements for hardware, software, licenses, networks, and data storage for the project. Always consider the future access of the model information. Software vendors usually upgrade their products annually, so make sure that any platform upgrades during the project don't cause any problems. In Chapter 7, we explore in more detail your software and hardware requirements.

At the center of your BIM process is the common data environment (CDE), which should be the main software priority. If this foundation isn't in place then failure is looking likely! Chapter 8 discusses the CDE in greater depth, but for now, remember that the CDE is a single place where information is brought together and shared. Think of it as the single source of truth.

With your implementation strategy well underway, you must take into consideration a few other things. In the next sections you consider who is the right person for the job and at which point you face the green light and go ahead and actually do it. Go on, you know you can.

Who is the best person to implement it?

Usually, the BIM manager takes the decision on software and hardware needs, in consultation with the whole team. Today many virtual desktop solutions exist, and most software is cloud based, which you need to factor into your considerations. Make sure that whatever solution you chose it's suitable for your interoperability needs and you have sufficient hardware power to make use of the information you create.

When is the right time to implement it?

You want to implement software and technology as soon as possible — the sooner, the better, and before the project commences. Make sure that you allow enough time to implement your IT strategy, especially if you're in a remote location. If you don't have the adequate infrastructure in your area, you may potentially require 4G cards or local caching servers, all of which will take additional time to put into place. Conceptualize the infrastructure and hardware solution to ensure the project team can share information both within and without your organization's firewall.

Your strategy may involve upgrades and software installations, so make sure you give the team plenty of warning that this will occur so that you can coordinate and plan IT improvements with minimal disruption to the day-to-day running of the office. Use the BIM execution plan as a way to document your IT strategy and communicate it to the whole project team.

Combining Information into a BIM Execution Plan

The *BEP* is essentially your digital method statement. You prepare this important and useful document at the pre-contract stage when you're tendering for a project. Following the contract award (fingers crossed that you won), you submit the plan post-contract. An overarching BEP may exist where more than one supplier has been appointed that contains individual BEPs prepared by each supplier.

The BEP is an important document within the supply chain's armory. Here are the main objectives of the BEP:

✔ Sets out the supplier's and supply chain's proposed approach to BIM implementation on a project

✔ Outlines your overall vision, along with how and when you'll implement BIM, allowing the whole project team to sing from the same hymn sheet

✔ Develops at an early stage of a project, and then is developed continually throughout the different project phases

✔ Defines scope and extent of BIM implementation on a project, including roles, responsibilities, levels of definition, and timings of information exchanges

✔ Is a living document that you should review regularly alongside model quality audits

You may want to develop your own BEP for a project in a particular office that may become quite standardized. A word of caution, however: Not all projects and clients are the same, so the BEP must respond and be relevant to the particular project in hand.

Chapter 8 discusses how the client describes his data needs within the employer's information requirements (EIR) document. The BEP is your way of proving to the client that you're up for the job and that you're the right organization with the right team. Within the BEP you get the opportunity to answer all those probing questions that the client sets out, such as when and how you'll go about the BIM project.

The BEP is considered in two distinct phases: First before a contract is awarded and then again after the client has awarded the contract to the successful bidder. The following sections consider the pre-contract BEP, including what this document should cover, and the post-contract BEP.

Knowing what to add to your pre-contract BEP

Include the following in your pre-contract BEP:

✔ **Overview:** Describe why you're creating the BEP in the first place. Imagine that someone is picking this document up for the first time and knows nothing about the project. Let him know what your mission statement is and give him an executive summary of the document.

✔ **Information required by the EIR:** The BEP should respond to the parts of the employer's information requirements that specifically require a response. This is likely to include information regarding

• Planning of work

• Coordination

- Clash detection

- Health and safety management

- Collaboration procedures such as details of meetings, schedules, and agendas

- Model management including filing storage, structure, and modeling requirements

✔ **Project information:** This section is usually an introduction for the other team members, and it helps them understand what the project is about as a whole. Include information such as project name, description, location, critical dates, key project members, project roles, a company organization chart, and a contractual tree. The project team uses critical schedule information for future reference as the project develops.

✔ **Project goals and objectives:** As well as documenting the strategic value and specific use for BIM in the project, include major milestones consistent with the project program.

✔ **Project implementation plan (PIP):** This document is submitted as part of the initial BEP by each organization that's bidding for a project. Include company organization charts and details of project roles.

✔ **Project information model (PIM) delivery strategy:** The major goals and objectives for the BIM implementation must be considered and stated as a project strategy document. This document should set out what tasks will be carried out and by whom.

Remember that not all digital data will be 3D; more than likely the project team needs to consider in the strategy lots of flat documents and how they link together, and so classifications and naming conventions are key.

Assessing the team

Putting together the right team from the outset is essential for the success of the project. When seeking others to be part of your team, consider the skillsets and knowledge that you require. The BEP is submitted by the supplier to the employer on behalf of the whole supply chain and includes a summary of the supply chain's capabilities and responsibilities. The PIP is a document the client can use to make a quick assessment and comparison of the people and organizations that he's intending to use, so make sure you sell yourself here.

The supplier submits the PIP as part of the initial BEP for each organization bidding for a project. You may find yourself asking a number of questions to potential partners or may even be asked if you're joining another team. These assessments are designed to look at your BIM competence and cover past BIM experience, IT, and resource.

Addressing your post-contract BEP

The BEP is a living document that you constantly update and review. Assume that you won the tender (congratulations) and now you must develop the pre-contract BEP further in the form of a post-contract BEP. The post-contract BEP should contain everything in the pre-contract document plus the following information:

- ✔ **Management:** Include details of roles and staffing, responsibilities, and authorities. You may not be able to complete this from the outset, but do update it as the project progresses and when the information becomes available. Also include any existing legacy data or survey information (if available) along with how you propose that information shall be approved.

- ✔ **Revised PIP information exchange:** This should contain up-to-date capability assessments that confirm the competence and capability of the supply chain.

- ✔ **Information exchanges:** Include what elements of the model are exchanged by discipline, when they'll be exchanged, and by whom. The responsibility matrix outlines the agreed responsibilities across the whole supply chain, task information delivery plans (TIDPs), and master information delivery plans (MIDPs). This document is key to success: Make sure it's comprehensive and well thought out with the right information at the right time.

- ✔ **Model QA procedures:** The supplier must define the overall strategy to ensure quality. Include details of modeling standards, naming conventions, model origin points, agreed tolerances for all disciplines, and change control procedures.

- ✔ **Delivery strategy and procurement:** They include details of the chosen procurement route and contractual details.

- ✔ **Communication strategy:** It includes details of how you intend to communicate with the project team, which mediums you'll use, and what types of meetings you'll have and when. Also establish protocols around how the project team will create, upload, circulate, and subsequently archive information.

- ✔ **Model structure:** Detail how the project team will structure the model; for example, how various disciplines break the model into appropriate schema, such as floor by floor. Also include file-naming structure and naming conventions, coordinate systems, classification structure, and modelling standards.

✔ **Facility management requirements:** Start with the end in mind. Engage facility managers at an early stage so that any specific requirements about the delivery and type of data are useful and meaningful in the post-construction phase. Understand how the transfer of data to the computer-aided facility management (CAFM) will work — can it be automated or mapped across?

✔ **IT, solutions and technology:** Examples include agreed software versions, exchange formats, and details of process and data management systems.

Putting everything together

Fundamentally, it's about ensuring that you and all project team members are using the same data and compatible systems, in order to realize the benefits of BIM process and technology, like better coordination and elimination of errors. It can be difficult to sum up your progress in a simple way.

The Capability Maturity Model (CMM) is a classic way to describe the development and sophistication of processes. When the US Department of Defense first developed the concept, it was with software development processes in mind. However, the resulting diagram of five steps can be used to demonstrate the progression of process optimization for many organizations. Governments and Fortune 500 companies alike have adopted it. It also forms the basis for the Information Technology Infrastructure Library (ITIL) process maturity assessment framework.

The diagram has real power when used to describe BIM implementation too. Recently, AECOM adapted the standard steps of the CMM to define its plans for BIM adoption, which Figure 13-1 shows. According to AECOM's diagram, the five steps to BIM implementation are as follows:

✔ **Initial:** You're aware of BIM and want to implement it, but you're still only beginning to understand where your organization needs to improve and can only effect change at a small scale, which leads to the next step.

✔ **Defined:** You're more capable of explaining the benefits BIM provides and are trying to apply BIM methodology to your organization at all levels through training. The process of applying what has been understood into real practice leads to the next step.

✔ **Managed:** You have embedded some BIM activities into real projects and are beginning to demonstrate the outcomes and return on investment. The implementation is joined up and is being coordinated by senior management as standard. This is the heart of BIM implementation. The process of embedding BIM processes into all project work leads to the next step.

✔ **Integrated:** You work in a fully BIM-ready organization. The protocols and procedures to enable advanced BIM to become part of the fabric of every office are well managed and measured. For most organizations, this is the realistic goal of BIM. The process of becoming fully operational with BIM at the center of best practice on every project leads to the next step.

✔ **Optimized:** This is an idealized stage of BIM implementation where all procedures have been refined to work at maximum efficiency and that leads to unprecedented levels of innovation across every office. Instead of best practice, this is *next practice,* generating high cost savings and exceptional building performance.

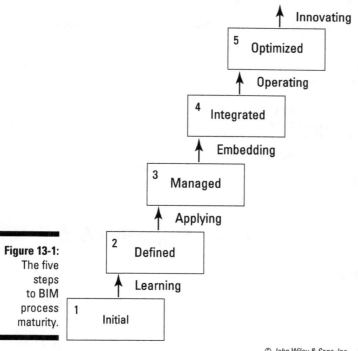

Figure 13-1:
The five
steps
to BIM
process
maturity.

© *John Wiley & Sons, Inc.*

Communicating with the team

Increasingly, organizations are producing BIM capability statements that give an overview of BIM experience and resources available. This is a good way to communicate with the outside world, but what about with your internal team?

You may be faced with explaining your strategy to the CEO or maybe to the intern. Whoever you're talking to, make sure the message is short, simple, memorable, and clear.

Inevitably, lots of questions will follow, and you may not be able to answer them all straight away. If that's the case, don't bluff your way through an answer. Instead, be honest and say that you don't know and that you'll find out and come back to the person.

Organizing information in an accessible way

People are producing more digital data than ever before. Social network platforms and cloud storage have changed the way in which people share information. However, in order to make sense of all that information, you need to organize it effectively so other members of the project team can produce, retrieve, understand, and archive it.

Having a companywide work breakdown structure (WBS) is important and useful. Protocols and quality management systems (QMS) in place may seem like extra work at the time, but they make office processes and quality assurance more efficient too. They can force you to tidy up your CAD block library into a BIM object library structure, to name your drawing layers consistently, or to put everything into centrally organized folders.

Modeling Your Plan after Successful BIM Processes

A number of templates and guidance documents can get you on the way to completing your BEP:

- ✔ The Construction Project Information Committee (CPIC) provides BEP templates for both pre- and post-contact as a free download at `www.cpic.org.uk/cpix/cpix-bim-execution-plan`.

- ✔ The Pennsylvania State University, Computer Integrated Construction (CIC) research program's BIM Planning website has developed the "BIM Project Execution Planning Guide," which is available as a free download at `http://bim.psu.edu`. The guide was a product of the buildingSMART alliance (bSa) Project BIM Project Execution Planning.

Now you are on your way, in the following sections we explain how to map the benefits of BIM, determine how much BIM costs, and make a profit on your investment.

Mapping BIM benefits

In order to map the benefits of BIM you need to have some sort of benchmark or data to inform your findings. Chapter 15 explores in greater depth the many benefits that BIM brings about, not just in the long term but also in the short term. Having the process is all very well, but you need to record the benefits.

However, using your in-house timesheet system (assuming that you already have one), you can look at the benefits on your productivity and bottom line. This is only one way to look at the benefits, but by using existing data that you're already collecting, you can start to see how more efficient a BIM process is to you and your organization.

Counting the cost of BIM

Coming to a formally agreed way in which you can measure benefits such as cost across the industry is difficult because no standard metric exists for measuring the return on investment for BIM. How do you calculate how much BIM has cost to implement? Do you include hardware and software? New salaries and consultant fees? The cost of this book? Do you record the per-hour savings in reworked drawings or meetings with the client? How much is the client prepared to pay for quality information?

In short, this is a hard calculation to quantify. Remember to include contractors and infrastructure people here too; this isn't just a building design type question. Consider the creation of a trial project delivery group that's agnostic and separate to the site team to measure benefits. Don't create lots of new key performance indicators (KPIs) — BIM should liberate value in your current drivers, such as money, safety, and program.

Making a return on investment

Making a return on investment, getting the figures to stack up, and ultimately generating a profit will be key considerations for most organizations. Construction generally hasn't adopted new processes like BIM as quickly as other industries, and often the reasons are financial.

To get the most out of your investment, make sure you

✔ **Learn from past mistakes.** Look at your past projects and ask why they were over-budget or perhaps not delivered on time. Try to look past the software to the root cause. Was there a communication problem? Did the client keep changing his mind? Were there some unforeseen ground problems? All these problems aren't exclusive to a BIM-enabled project, and apply equally to CAD and hand-drawn projects, but you can use the data within a BIM project to evaluate its effectiveness. Sometimes admitting where things went wrong is hard, but focus on the BIM process and work flow and where you can be more efficient and improve on your return on investment.

✔ **Change your process.** Integrating BIM into your existing or current work flows gives you a return on investment, but to really reap the rewards, look toward a new process for project delivery, one that uses the principles of collaboration, coordination, and trust. Consider investing in tools designed to increase collaboration, such as installing a dedicated BIM room for model viewing. Some of the most advanced versions can turn formerly dull project meetings into immersive 3D experiences, more like seeing a 3D movie at the cinema. For a great example of this, see the Texas A&M University BIM Computer-Aided Virtual Environment (CAVE) at www.youtube.com/watch?v=ErjvILsxvgE&feature=youtu.be.

✔ **Compare other similar projects.** Comparing apples with apples is a good way to calculate return on investment. Look to where a project may share similarities such as size, location, type, and materials with another project; for example, you can focus on the differences between a CAD approach and a BIM approach in terms of work hours and effort.

✔ **Look at the bigger picture.** When looking at your fees, you may find that you're going over-budget during the design phases. Because more effort goes into the earlier stages of a project, a BIM process produces much of the groundwork in producing documentation for the construction, so make sure you factor this in.

Chapter 14

Considering the Issues that BIM Presents

*B*IM is a game changer. The digital revolution has been slowly building for decades, and the industry badly needs BIM to bring it into the 21st century, ready for a connected, mobile world. BIM implementation and practice is demonstrating that there's also lots of room to grow and potential for even greater innovation and bigger savings. BIM helps to make construction of buildings and infrastructure less fragmented, breaking down silos and increasing communication and connection.

However, thinking that all this new, exciting BIM stuff comes risk free would be foolish. In fact, you need to think about a number of issues when considering BIM projects and disruptive processes that can impact your business. We show you that, in most cases, you can overcome the challenges and difficulties you encounter just by thinking about things in another way or working differently.

This chapter helps you to think about how BIM can change your working methods and business activities and takes a look at the cyber security risks to do with where your data is hosted and the many legal complications that arise from introducing BIM into an industry with existing contractual arrangements. You need to consider who owns the information in the model and what party should be blamed when something goes wrong. You need to be able to assess risk and calculate fees or insurance to match. Finally, we explain how to make sure you have the best chance of making BIM into a worthwhile investment for your business and everyone you work with.

Managing Change in Your Business

In Chapter 6, we ask you to name the processes that you thought made your business less efficient. Our intention is to help you to plan ahead for BIM implementation. After you're in the thick of introducing new platforms and practical work, you'll encounter different issues that result from your actual projects and the way that day-to-day work can put limits on your original vision for BIM.

Determine what the processes are that you keep you from reaching your vision for a BIM-driven company. (Hands up if you're thinking of some *people* too!) Carefully and logically, you can begin to implement the changes you want to make. This process of organizing measured and planned change is called *change management*.

Change management is the approach you take to getting your team from where it is now to where you want it to be. Because BIM is a process, change management is a crucial part of successful implementation. In order for anything to evolve, two things need to be in place:

- ✔ **Reason to change:** This can also be called *pressure to change*. It's the incentives or obstacles that eventually force an organization to implement change. A good example would be an update to legal regulations to which you have to adhere. In that situation, you have no choice but to change your processes to match the new legislation.

- ✔ **Capacity to change:** In this sense, *capacity* just means how busy your organization is and how workload could prevent individuals and teams from having the time, patience, or energy to actively participate in changing processes.

The following sections examine some key factors that can block adoption of new systems and how you can support colleagues who are holding things up, whether it's intentional or not. We provide you with some great tips for dealing with BIM doubters too.

Forgetting the baggage

People often fail to adopt new technology and systems effectively for some simple reasons. Here's a list of three factors:

- ✔ **Lack of change management:** If you're in charge of BIM implementation, make sure that you clearly understand the needs of the users who will have to use these tools on live project work every day and prepare them for change. As a manager, you may need to answer queries about very practical matters, such as office layouts or job roles as a result of promoting BIM. Be prepared to respond to talk about what BIM does and doesn't mean for your business. Take a proactive approach to supporting the concerns of colleagues before they have the chance to impact others. Remember to be consistent — a delayed date for meetings can damage relationships.

✔ **Lack of training:** Your implementation of BIM processes can fail because end users don't have the right level of support and training. Sometimes this is practical training about using software platforms, but you may also need to train colleagues about your BIM strategy. You understand your aims and objectives for BIM implementation and are confident in the tools and processes you use, but others may not share your enthusiasm. Realize that everyone has individual skills and strengths. Some individuals thrive on technology, data management, or object modeling, whereas others may struggle. Some may be reluctant to change for reasons of tradition, but bear in mind that some may have a fear of failure or of being found out in not understanding the tools or the vision. BIM needs to be an enabler, making sure people can continue to add value and contribute their expertise to the larger model, not forcing them to rethink their career choice.

✔ **Lack of quality:** Confirm that the products and services that you're advocating are of a high quality. Users are instantly frustrated when unreliable hardware or malfunctioning automatic services get in the way of work and delivery to the client. Coordinate your plan with IT infrastructure teams to ensure that your internal systems are capable of supporting new technology, from Internet speeds to the specification of individual computers. Equally, be aware that when systems appear to be taking longer to use than traditional processes, it may be due to the learning process, particularly for complex software or procedures.

Dealing with cynical people

In trying to talk about BIM, you're likely to meet with some cynicism, because doubt in the face of change is part of every industry. You need to do your best to win people over, because implementing technology changes is much easier when everyone is on the same page and understands how they benefit from the change.

Some people may say or act like they've heard it all before and say that when CAD was introduced everyone talked it up in the same way as BIM today, but that it didn't transform the industry. Demonstrate that this isn't a fair comparison by showing that CAD was a format change, like vinyl record to CD. Think of how CAD used the concept of *paper space* to mimic physical drawing methods. Instead, BIM embeds intelligent objects with data and provides a huge leap forward in terms of information exchange. Keep encouraging people to separate BIM and computerization in their minds.

Paul once had a conversation about BIM with someone at a conference, who then pulled a pen out of his jacket pocket in order to prove that he still sketched designs by hand and that "computers would never replace humans." He definitely had missed the point!

You can demonstrate value through case studies and success stories, especially about businesses of a similar size to your own. People are more likely to adopt technology if they see adoption by their peers. Academics call this the *contiguous user bandwagon* — basically, no one wants to be first to make the jump.

Choose case studies of an appropriate scale. Make your example too small in comparison to your own business and the increased agility that comes from a small team probably played a role in BIM success; make the example too large and the increased access to resources and finance is likely to skew the result in comparison to what you can achieve.

Digital technology has impacted the construction industry, which continues to evolve toward a fully digital future. However, you may encounter resistance to and cynicism about BIM and its related topics.

Here are some useful ways to communicate the benefits of BIM and advocate a modern, digital construction industry to your colleagues:

- ✔ **Use personal experience.** Rely on something familiar to describe how other technology has disrupted and impacted people's daily lives. You can find lots more on digital technology in Chapter 6. Ask the person when she last used a paper map or a payphone. Demonstrate the real-term benefits. You're not just experimenting with science fiction; this is the future of the workplace.

- ✔ **Find an example of connected data.** BIM is full of layered data, so using a comparison for this kind of connected information is helpful. For example, take an online/GPS location-based tool for a comparison. You use it instead of a traditional map or road atlas, and can produce street-by-street directions from your current location to a destination, or plan a trip to any location on the globe in advance, from the huge database of map information.

- ✔ **Provide an analogy for information modeling.** Because of all the layered data in BIM, you need tools to search and interrogate the information. Sticking with the map example, explain that the directions are just one filtered view of all the information available in the online/GPS location-based tool. You can then choose to interrogate that information more closely by finding images of the actual roads you'll use or searching for local businesses such as restaurants. You can layer additional content such as traffic information onto the map in order to take the quickest route, see public transport links, or use GPS data from friends' devices to locate them at your destination. This is modeling information — pretty amazing.

✔ **Compare construction to other industries.** Show that every industry undergoes digital transformation, so it's understandable that the construction industry has seen the evolution of online tools and CAD over the years. Demonstrate that project teams still carry out the vast majority of building design and construction using the same analog methods and processes that were used thousands of years ago — paper drawings, printed documents, standard materials, and hand tools for site work. Compared to most industries, construction remains significantly behind in terms of disruptive, major change to its traditional background. Point out that plenty of examples that push the boundaries now exist, but that the progression is very gradual.

✔ **Combine information modeling and building processes.** Imagine that you can combine the power of online mapping tools, the intelligent geometric data of CAD systems, and the ability to interrogate the information model of a building project to find whatever information you're looking for, from the airflow rate of a duct to a manufacturer's address or the replacement cost of a component. This is Building Information Modeling.

✔ **Extend BIM across the project timeline.** With BIM, you can interrogate the model to show every instance and every stage of the product, including its embedded data (such as a warranty expiration date). Demonstrate that from the earliest design brief, the client can see the project design in great detail. Design and construction teams can coordinate data to reduce errors and ensure fewer changes are required on-site where costs increase. Then ask your colleague to consider how a facilities manager currently finds out the recurrence of a product that needs replacing. The manager probably has to trawl through various boxes of documents and drawings to find every instance.

Prepare positive and effective training that will support users not only to understand BIM as a theory but also through the inevitable changes to working practice and processes. You can begin with templates and framework course content, but tailor everything to reflect your organization's unique structure, projects, or situation.

Keep the long term in mind. Focus on the essential processes and functions that will impact you from day one, because you need to measure return on investment, but don't lose sight of potential development, both in technology terms and in your own business. For example, if the software or services you consider have additional features you won't use, take the time to think about why that is and how you may add them to your future processes.

Acquiring and retaining corporate knowledge

One of the most important things to ensure when an existing process needs to change is that nothing of value will be lost. The most useful procedures are often little things that have helped your business to get where it is today. Look at your existing process and see how valuable aspects can be used to enhance your new procedures, instead of discarding them.

In particular, those individuals who've clocked up the most years often have the most expertise and embedded knowledge. It may be in technical skills, real-world project experience, or just the kind of life experience that means they know where problems occur and how to fix them. But where does this knowledge live? If it's in one person's head, then it disappears as soon as that person leaves or, more often than not, when that person retires. As quickly as possible, begin the retention of corporate knowledge (sometimes called RoCK), where through office guidance documentation, continuing professional development training sessions, mentoring, and plain old better communication, you keep hold of the expertise long after individuals have left the office.

Here's an example. Say that you work with Bob and he's your go-to guy when you need a technical detail of a complex junction or structural connection produced. But Bob finds it tough enough to move from paper to CAD processes, and so he likes working in 2D paper-space and putting all the lines on one layer. He's a few years from retirement and you want him to become part of your BIM process. Consider why Bob does what he does and why it's been so successful in the past. Here's a list of things that may seem like negatives, with another way to look at them:

✔ Bob is used to drawing on paper, which means he's likely to invest care in his work and pay good attention to detail. Because of how long it takes to correct errors, paper workers are often more measured in their decisions and less likely to make mistakes. You can use staff experience to improve methodology and approval in digital working environments. People who are used to working on paper are generally brilliant at setting up revision logic.

✔ Bob needed a lot of support, which means he'll understand the very issues colleagues are struggling with. Perhaps the reason Bob is trying to replicate paper processes in CAD without really using the power of the tools is insufficient training. Often those people who needed the most training at the beginning can become very good trainers in the future. Helping everyone to understand that BIM is an overarching process and not just about the software means that anyone can play a key role in its implementation. Training is fundamental to BIM success.

Try to understand why traditional processes feel natural. Familiarity plays a part, but perhaps Bob just needs a coaching mentor or simply someone he can ask questions.

✔ Bob has very specialist knowledge, which means that he knows things that no one else in the office does. Get Bob to write some office guidance on his areas of expertise. In this example, because junctions and connections can be one of the trickiest parts of modeling the built environment digitally, that traditional craft can become a fundamental principle for future modeling. Encourage the creation of a new role, specifically for knowledge management and sharing.

BIM tools and platforms will never be a substitute for quality work and technical expertise. They'll draw attention to both quality and errors.

Many people believe that BIM makes their job obsolete and they're concerned about redundancy. Think about how you'd feel if your particular specialty were hand-drawn perspectives that took many hours of setup, artistry, and skill to produce. Make sure you understand what everyone can offer. In all your positivity about what BIM can do for your business, don't forget to explain things at a personal level with one-to-one sessions.

Clarifying the Legal Implications of BIM

If you already have experience as part of the built environment industries, more than likely you've encountered the regular collision of construction and law. From procurement routes that try to allocate risk to lawsuits and disputes when things go wrong, building law has been evolving for a long time. As you become more familiar with BIM, you quickly run into questions about the legal aspects of these new processes. Trying to implement BIM results in a lot of questions about contracts and law. The key to BIM project success is collaboration, but you may need a legal framework to determine the rules of the game.

We should point out that although we all have experience of project contracts and the cases we know, none of us are lawyers. The following sections raise some interesting points about the legal implications of BIM, although we don't intend it to be legal advice. The extent of liability and risk can be a real worry for many businesses. Here we list some key issues you may need to think about, and we recommend that you consider contacting specialist legal professionals for advice if anything concerns you.

How many people do you work with, side by side, to produce a typical project? You can probably hazard a pretty good guess at that answer, so we reword the question slightly: How many people use the information you produce on a typical project? That's a lot more difficult to answer, right? At any given time, clients, designers, contractors, subcontractors, fabricators, builders, city officials, manufacturers, owners, and janitors could be using something you generated.

In the majority of BIM projects, your information needs to be read, exchanged, and interrogated by other people throughout the process. In what we refer to as a Level 2 BIM world, that's done by federating information together, everyone contributing to a central data store. In a Level 3 world, this is forecast to happen in a central environment where everyone's working on one source of the truth.

The latest figures suggest that the industry wastes nearly $15 billion per year through interoperability issues. In this context, *interoperability issues* means the problems associated with trying to work with information in different formats, from different teams, in different locations, and turn them into usable data to develop into your own work. BIM attempts to solve many of these issues through common information exchange formats and data sharing. The idea of a federated model is to bring together everyone's data to avoid duplicated effort and understand where clashes occur. But what's to stop someone using your content on another project and passing it off as her own? Do the same rules apply as in a paper world? Does a client have the right to pick up a fully designed building model and replicate it in many global locations?

Facing where your data lives — digital security

One of the first issues to consider is data security. This cultural mentality is quite familiar to big business and finance, but is still relatively alien to both construction and legal industries. In simple terms, you need to ensure that controls are in place for the secure, confidential, and protected storage and exchange of information.

General opinion is moving away from a traditional idea that you keep your business or project information in filing cabinets in the office. A lot of the content generated in people's daily lives is now hosted in the cloud, in remote data centers around the world. Previously, people may have considered business data too sensitive for centralized data stores, but that's no longer the case. Modern data centers and IT infrastructure providers often guarantee the security of your business-sensitive content held within.

Not just that, but building management systems (BMS) or building automation systems (BAS) contains significant data about the operation of buildings. Some data is used to control alarm and security systems, so if the system were to be compromised, the whole facility's security could potentially be breached. The more connected smart built assets become in the future, with the Internet of Everything linking all sorts of things online, the greater the risk that security breaches will no longer be isolated to just digital data but to physical buildings too. Therefore, security policies have to treat data breaches as part of the overall security of a built asset.

As part of the UK suite of documents designed to describe Level 2 BIM, PAS 1192-5 provides guidance on cyber security. The UK BIM Task Group describes the document as outlining:

> *Security threats to the use of information during asset conception, procurement, design, construction, operation, and disposal . . . the steps required to create and cultivate an appropriate security mind-set and the secure culture necessary to enable business to unlock new and more efficient processes and collaborative ways of working.*

In theory, you shouldn't think of PAS 1192-5 as just BIM guidance, because it will apply to any built asset that stores digital information. You need to check that relevant security issues for a particular project or client are considered and documented in the employer's information requirements (EIR). For example, do you have access to other collaborators' confidential information?

 One of the side effects of highlighting the risk of BIM data security is that everyone's insurance premiums will go up. In reality, standard insurance policies for most built assets simply aren't designed to cover information breaches like this. The types of policies BIM projects potentially may need have previously been reserved for big business, finance, and data storage.

Controlling liability and risk

Asking who's responsible when something goes wrong is logical. In traditionally procured projects, you probably have quite clear opinions on responsibility and the balance of risk. Most contracts are trying to come to an arrangement where risk and reward are shared to everyone's satisfaction. Think of a contract as a seesaw between profit and risk.

When disputes and problems arise in a project, you need to refer to the contract. If disagreements continue, then it's up to third parties and courts to determine how to interpret the contract and how to apply the law to individual situations.

The benefits of BIM (like better data) should actually reduce the number of disputes that end up in court as a result of poor project information. However, BIM adds another layer of complexity to assigning risk and the balance of liability. Think of a BIM process across a project timeline; all parties need to agree on liability before you start to collaborate and share your information. The risk is that BIM leads to a different set of disputes.

In particular, you need to understand who has liability when the various information from different parties is brought together. As we talk about in the section "Dealing with intellectual property and ownership" later in the chapter, *ownership* of federated content is clear because it remains with the creator. However, ownership becomes hazier if someone in a BIM coordination or BIM management role changes the federated model. If a deletion, edit, or omission in the model management process causes a defect in the built asset, who's at fault needs to be clear.

For example, the US ConsensusDOCS BIM Addendum limits each participant's liability to reflect her discrete responsibilities and existing appointment relationships. Whatever contracts, protocols, and document standards you use at the outset of a project, team members mustn't use collaborative BIM as an excuse to assign unreasonable liability to parties of significantly different scales or responsibility.

It gets much trickier when you move toward collaborative Level 3 BIM, where the aim is that all parties use only one model of the built asset. The collaboration and merging of digital content is likely to make assigning blame for mistakes much more difficult. You need to ensure that your contract clearly defines design responsibility and your own liability for the project at every stage of its development. The question is whether eventually you'd be happy to share liability for the work of all the team across the entire information model and trust that everyone is working to the same high levels of quality and integrity.

Most people we've talked to hadn't considered the risk associated with software errors. Software vendors usually protect themselves from liability with terms and conditions that basically say you use their platforms at your own risk and the requirement is on you to check design, federation, or detail, and not just assume the software is doing things right.

Check that your contracts reflect the limited liability of software vendors by clearly setting out responsibility for software selection and expertise. For instance, if projects are mandating the use of particular platforms or versions, in the event that software fails, the contract should pass on liability to the party that required the software.

As a designer, consultant, or contractor, you're assumed to have carried out the necessary due diligence on the methods, tools, and platforms you use. Verify that you have insurance that protects you from mistakes in your outputs. The likelihood is that more shared BIM insurance products will become available to protect multiple participants from the risks of collaborative working and the potential errors of model management. Like everyone else, the insurance industry is playing catch-up.

Dealing with intellectual property and ownership

Another issue about BIM you may have heard about is who owns the model. Thankfully, copyright and intellectual property is a bit more straightforward than contractual liability. You want to be sure that you have just as much ownership of your intellectual property as you did before BIM implementation on the projects you do.

Ownership of the information should always remain with the creator. Federation makes keeping ownership with the creator easier by drawing distinct virtual lines between the content of one user and another. These portions of the model don't merge with one another, so you know where each bit of data has come from.

Model ownership isn't entirely faultless, though. Think about how you may design placeholder objects and eventually substitute these with manufacturer or subcontractor objects. That piece of design has two owners, the copyright invested in the object, which is the manufacturer's intellectual property, and your design decision to use it in a certain way.

Collaborative working makes drawing those virtual boundaries more difficult, especially at a design stage. For example, a landscape architect and contractor may work together to consider site drainage, and you could easily consider the planned solution as the design work of both parties.

Another great benefit of modern BIM platforms is the ability to track changes to the BIM and quickly determine who made a change and when in the process. Engineering industries like aeronautics are much better at this kind of audit trail than construction. The trail proves very useful when something goes wrong, because you can see who made design and construction decisions along the way.

A different way of looking at the situation is that the various project team members now have much greater access to your design process than ever before. Instead of just providing a set of outputs, be aware that there's nowhere to hide when it comes to showing your working methods. Opening up work in progress to scrutiny increases overall quality across projects, and ideally rids the industry of cowboys, but you may want to adapt your methods to this new level of exposure.

On projects, either the client has a license to use the content of the model for the one-time-only purpose of construction and facility management, or each party allows everyone else to use their contributions for a discrete length of project work. Generally, a client will appoint someone to manage sublicensing of content for other parties on the project.

Then the contract must make it very clear how much ownership of the design the client has after the project team completes the information model. If the client wants to use the model for asset management purposes, then that's reasonable. In fact, the provision of a digital version of the built asset as well as the physical one is a key benefit of BIM. However, a client replicating your contribution in a completely separate project a few months later isn't reasonable.

Arranging contracts and expressing requirements

Making the guidance in all the preceding sections, like that on liability and intellectual property arrangements, work relies on the right contracts. The history of BIM has made it clear that most construction contracts at least need some adaptation, if not brand new terms, and existing industry institutions will design some contracts completely to support BIM processes. Legal organizations and a few BIM groups are developing their own.

You can use a BIM protocol, like the Construction Industry Council BIM Protocol (www.bimtaskgroup.org/bim-protocol), as the supplementary reference in the contract. Check that the relevant protocol explains the collaborative environment of BIM and the game's rules. Support and cross-reference through the BIM execution plan and appoint an information manager to oversee this.

One of the most important parts of the contract determines whether the BIM itself becomes a contract deliverable or the single source of truth from which the project team generates contract deliverables (like drawings or specifications). The client needs *data drops,* which are extracted snapshots of information from the model at various stages, so that stakeholders can determine whether the project proceeds, based on cost or design criteria. As well as the regular data drops, you may also need to deliver the completed model as part of the project package.

A lot of talk about BIM contracts will mention integrated project delivery (IPD). IPD is a different strategy for sharing risk and liability, where the client, lead designer, and contractor all enter into the same contract, instead of each party having individual contract arrangements that might contain conflicting information. Advocates of IPD say that because risk is shared, the method improves the quality of the finished product. You can also use this method of procurement to get early contractor involvement, which has numerous project benefits.

The legal profession and the construction industry haven't always gotten along, but now everyone must communicate better than ever about BIM and new technology, working in cooperative ways. The protection of law should encourage your involvement in BIM, not make you feel that the risks are too high to get involved.

Ensuring Your Business Thrives with BIM: Engineering Collaboration

You want BIM to succeed in your organization and for your projects to reflect the benefits of working with these new methods and new technologies. One of the most important aspects of BIM is collaboration, and working with other teams, both internal and external to your organization, can present many challenges you'll need to overcome.

Along with the change management and contractual issues covered in this chapter, here are a few other issues around collaborative working you might want to consider:

- ✔ **Working together using different software:** Don't forget that many different software vendors exist with a range of platforms and products. In Chapter 11, we mention ways that the industry is looking to output *open BIM,* exchangeable formats like IFC and BCF that can be read by most software platforms (with various levels of success). However, the best way of ensuring that you can work with everyone effectively is to agree on the platforms you're using at the outset of the project.

✔ **Avoiding ability misunderstandings:** Some people believe they're doing fully collaborative Level 3 BIM already, when in reality doing so is almost impossible because the technology doesn't exist yet. Don't be afraid to make the facts clear to your clients, stakeholders, or managers. You may feel you're being honest about your capabilities, but losing out to organizations that are claiming more than they can deliver. A lot of BIM wash exists across the industry, but people are gradually being found out.

✔ **Considering who needs to understand your BIM plans:** Don't just look at the project team when considering your options, but also think about the various additional teams that the changes may affect or that need support. Remember that implementation and setup come at a cost, along with annual running costs, and that other teams need to be aware of the purchasing costs. For example, how will you explain BIM to your IT provider or your risk-management consultant or insurer? Who else needs to understand BIM?

More than likely, you've heard the phrase "change is inevitable." Change is fundamental to moving the construction industry forward, whether because you're being pushed by government policy to adopt new processes or because you can already see the financial and quality benefits for yourself.

Quickly and clearly communicate the reasons for change to your colleagues and the areas you have responsibility for. By using change management techniques, you can reduce the negative impact of change on your business, along with speeding up the acceptance of change and recognizing the rewards.

BIM can do lots of things for your business, but one of the essentials is that it improves the quality of information throughout your projects, offices, and administration, and increases your reputation for great work. If you have existing quality assurance processes in place, BIM implementation should work hand in hand with these to ensure that the project team can easily retrieve clear, accurate data at the right time, and print, export, or transfer the data as high-quality outputs.

Be aware that some areas of the industry will resist BIM because its efficiencies and effects threaten their existing commercial business. You can automatically take off and analyze quantities from the information model or repeatedly refine facility management information during the design process, and you directly impact established businesses that provide these traditional or manual services. Notice that the speed of development is increasing, and so planning ahead for the impact is hard. Every aspect of the industry needs to evolve to survive and thrive, including you.

Part IV

Measuring the Real-World Benefits of BIM

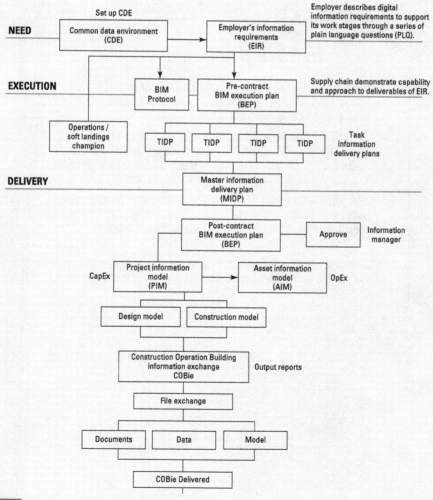

NEED

Set up CDE

Common data environment (CDE)

Employer's information requirements (EIR)

Employer describes digital information requirements to support its work stages through a series of plain language questions (PLQ).

EXECUTION

BIM Protocol

Pre-contract BIM execution plan (BEP)

Supply chain demonstrate capability and approach to deliverables of EIR.

Operations / soft landings champion

TIDP   TIDP   TIDP   TIDP

Task information delivery plans

DELIVERY

Master information delivery plan (MIDP)

Post-contract BIM execution plan (BEP)

Approve

Information manager

CapEx

Project information model (PIM)

Asset information model (AIM)

OpEx

Design model   Construction model

Construction Operation Building information exchange COBie

Output reports

File exchange

Documents   Data   Model

COBie Delivered

John Wiley & Sons, Inc.

web extras

Check out how BIM can make construction safer for everyone involved at www.dummies.com/extras/bim.

In this part . . .

✔ Evaluate the actual impact of BIM on real construction work and see examples of great case-study projects where the benefits of and investment in BIM have been proven.

✔ Reduce risks on-site and during building maintenance with BIM by making the right decisions early in the design process, and make the construction industry not only a safer place in which to work but an industry of which you can be proud.

✔ Strengthen the entire BIM supply chain by understanding how everyone involved uses your information differently — for different purposes and at different points along the project timeline.

✔ Discover new ways in which facility managers can use information to maintain and run the asset to its full potential.

Chapter 15

Evaluating BIM Benefits and Investment

*O*ne of the first things a company is likely to ask when evaluating business benefits that BIM brings and the investment required is: How much will BIM return on my investment and how much money will it save or even generate? What about carbon benefits and new efficiencies? Think about your traditional processes, drawings, paper printing, photocopying, postage, and so forth. What can you do to create immediate and tangible efficiencies? Are there good examples of projects that have shown the benefits of BIM that exist today, especially energy savings? Not just the future stuff about interrogating a spinning hologram, but a practical, grounded set of key benefits available to clients and investors today?

In this chapter, you discover that other industries aside from construction are going through a digital revolution. You explore how it's impacting construction and how you can measure the benefits and impacts to you and your organization.

Arriving at a Digital Understanding of Construction

When you look at the world around you, you see that everything is going digital. (Refer to the nearby sidebar to see how digital has affected so many parts of today's world.) The digital age also has finally caught up with the construction industry. BIM is moving the industry from a document to an integrated database way of working.

Think about the design process, which has seen the demise of the drawing board, giving way to the computer and graphics tablet. Today, architects and engineers are modeling buildings rather than drawing them, and the implications are profound and result in considerable changes in the approach. Traditionally, the project team produced and viewed together 2D drawings to imply a 3D representation. Now, 2D drawings are an output of a 3D model; architects and engineers generate them relatively easily using computer graphics, allowing the construction industry to produce the design and construction documents more quickly. Then client, end user, and facility managers can make more effective design decisions at an earlier point in the design process, which means that the project team has greater opportunities for coordination, collaboration, and production efficiencies.

Digital construction is a combination of innovative people, digital tools, and integrated processes, all of which are underpinned by information. Today's clients aren't just procuring built physical assets, but they're also procuring digital information that can help them make considered choices about capital investment and about the future maintenance and operation of that asset.

BIM is a cultural and behavioral change program above all else. Be mindful that some members of your team may need more assistance and training than others to get them up off the ground.

How the world has advanced digitally

Many other industries are benefiting from advancing digital technologies. The construction industry should be outward looking to other industries and learn how they have gone about it and what effect this has had on their industries. Here are some examples to show how far the world has come since Guglielmo Marconi and Karl Ferdinand Braun were awarded the Nobel Prize for physics in 1909 for their contribution to wireless telegraphy.

✔ Today people's auditory senses are listening to crystal-clear digital audio broadcasting music on the radio.

✔ People's visual senses are treated to digitally processed transmissions with the advent of digital television as television moves away from analog.

✔ While the old school among you hold on to your vinyl LPs for dear life, you can't ignore the arrival of music download sites such as iTunes and music-streaming services such as Spotify.

✔ Video cassettes are virtually non existent. Gone are the days of lining your selves with VHS tapes in covers to resemble books. You now stream and download movies to your hard disk (probably in high definition and stereo digital surround sound).

It's fair to say that a digital revolution is in full swing. Sites such as YouTube, Facebook, Instagram, and Twitter are changing the way people communicate and share information with one another, and people are producing digital information more than ever. However, to make sense of the data and to make it useful, you need structure and consistency. Structured and consistent data is where the real potential of digital information lies.

You only have to look at other industries, such as photography, music, and automotive, to acknowledge that the world is moving from analog technologies to digital. This means that to keep updated with advancing technologies, the construction industry needs to reevaluate its current processes and procedures. In the next sections we discuss how new technology is changing the industry and how accurate information can help with technology.

Advancing standard processes with new technology

The use of technology and defined collaborative processes brings about greater collaboration between project teams, which in turn leads to better project deliverables through improved quality and accuracy and flow of information and data.

Before you retool or reengineer processes, you need to establish benchmarks against which you can evaluate tangible items such as cost, time, and carbon, while being careful not to create new layers of key performance indicators. You can measure the things BIM can generate, but make sure that you also consider measuring the things that BIM stops happening, such as clashed reduction when requesting information.

Understand that you rarely achieve savings from BIM in isolation, but rather when BIM is part of a much wider program. As you discover in Chapter 9, BIM is one part of the pie that also includes slices of new methods of procurement and soft landings. Figure 15-1 describes a best-practice, collaborative workflow for a BIM supply chain.

A lot of acronyms are involved and we explain them in more detail in Chapter 8, but here's a quick overview. You start the process with an assessment of the overall need and the client records his digital information requirements, which form part of the contract documents. The supply chain responds via BIM execution plans (BEP) to demonstrate their capability to deliver the client's requests. Within the BEP, the project teams develop individual task information delivery plans (TIDP) that the main contractor uses to manage the master information delivery plan (MIDP). The benefit of the common data environment (CDE) is to provide a single location for the domain federated models and project data.

Don't rush out and buy software and hardware first. Your doctor wouldn't prescribe you medicine without first diagnosing what was wrong with you. Many different vendors and products are available on the market. Some are on a pay-per-seat license, and some have a concurrent or network license. Some software is subscription based and other software may be free. As you discover in Chapter 7, you need to consider a number of things before making any purchase or investment.

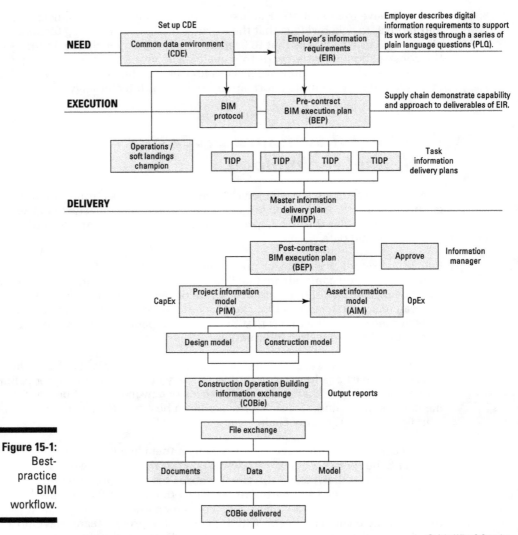

Figure 15-1:
Best-
practice
BIM
workflow.

After you establish your benchmarks, start reworking processes to become lean. Next look to remove any bottlenecks. Don't think only in technology terms, though, or software and hardware that may be holding you back. Also consider people and work on any barriers that may exist, such as training needs, fear, or resistance to change. After you do all this, then you can begin to marry digital tools in order to implement BIM.

Make sure you understand your client's requirements. Through the use of plain language questions (PLQs) the client is able to both set out and identify the minimum information to exchange at any given stage in order to answer the client's queries.

Ask yourself: how can BIM help automate some of your existing processes? If you manage data in a consistent way during the design process then you can pass data on for manufacture and assembly (DFMA), which is all about designing products while keeping the manufacturing and assembly process in mind. Through a DFMA process, you can make further savings.

Often, the reported benefits of BIM are directed toward the industry as a whole or individuals rather than organizations. The issue around how long your return on investment will take is an important consideration. You need to set aside adequate time for meeting training needs, reengineering processes, and adapting to new ways of working. Refer to case studies and talk to other similar organizations. In Chapter 22 we discuss a number of online communities in which people are always willing to share experiences and advice.

Processes should be outcome based. Therefore, consider goal-directed planning.

Reporting on the right metrics

In the UK, the Government Construction Strategy has been clear in its 15 to 20 percent target for costs savings, where these cost savings are from the construction budget and not post-occupancy costs. The benefits to industry have been presented by performance improvement, greater project certainty, and a reduction in risk, which are all factors that an integrated supply chain can quantify and understand.

To achieve the benefits outlined by the UK government and make intelligent and best use of information, you require a good flow of accurate information throughout the project teams. The big value proposition for BIM is in the operation and maintenance phases of a project. This information can be used to achieve optimized performance, running costs, and energy and carbon reductions. BIM brings great benefits to the design process, where the client can quickly test and accept or dismiss different opportunities and configurations to achieve optimal deliverables in terms of cost, quality, and time.

In the early days, the benefits of BIM were based on a leap of faith. Since then the construction industry has assessed the benefits across the world, through trial projects, cases studies, and research.

Assessing Cookham Wood

An interesting case study to look at when evaluating the benefits of BIM is HMYOI Cookham Wood Young Offenders' Institute. In testing the UK government's hypothesis for Level 2 BIM, it selected a number of early adopter trial projects. The Ministry of Justice (MOJ) has been implementing collaborative forms of procurement and early contractor engagement, and it recently brought BIM into the mix. Rather than carry out costly repairs, the MOJ decided to rebuild part of the prison, comprising a new 179-room accommodation block and associated education facility for the Youth Justice Board. This project was an ideal opportunity to test the objectives of UK Government Construction Strategy initiatives such as soft landings, Lean delivery, and BIM.

The UK BIM Task Group has been open and transparent about the process and published its experiences. A trial project delivery group helped validate the benefits, highlighting 34 lessons learned along with corresponding actions, divided into pre-tender and post-tender periods. The lessons went a long way in informing further work that the UK government needed to implement, including robust client information requirements that lead in part to the formation of the NBS BIM toolkit. Clients' informational requirements must state that the project team should link data to the model and output data in a format that the client can test electronically.

This project not only proved that BIM is a better process by which to work, but also demonstrated 20 percent savings achieved through collaboration with the supply chain, leading to innovative proposals and early clarity on the outcomes. The following figure shows a view of the federated model, combining the discipline-specific domain models; for example, architectural and services content.

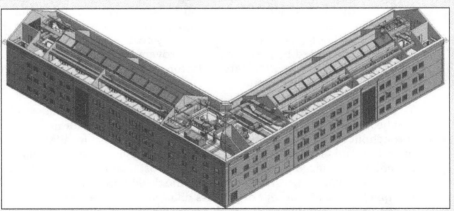

Illustration by The BIM Task Group

The UK Department of Trade and Industry set up Avanti projects (`http://constructingexcellence.org.uk/resources/avanti`) in 2002 to promote ICT-enabled collaborative working. Rather than a particular project, Avanti is an approach based on getting people to work together, providing processes to enable collaboration and applying tools to support collaborative working. A number of case studies demonstrate significant savings achieved by using BIM techniques.

BIM isn't a silver bullet to solve every potential cost issue on a project. However, it does go a long way in improving collaborations and the quality of information, which in turn makes for better informed decisions, project deliverables, and predictability. Through the use of frameworks, protocols, and standards, you can rectify many potentially costly mistakes at an early stage before construction commences and before a spade hits the ground. At this point, any design changes are relatively low cost because you're only moving around components within a digital model, as opposed to physical walls in the built asset (see Chapter 8 for more information).

BIM can give you more than cost certainty and savings, though. You can begin to build smarter and faster and be kinder to the environment. You can use project information throughout the whole timeline, through into the occupancy and maintenance stages, where you achieve further cost and carbon savings.

As we discuss in Chapter 17, BIM can be effective in making safer assets, not just during construction but also in occupation. BIM brings about a number of commercial advantages, but the potential benefits of improved safety by reducing accidents and fatalities are so high that they're almost unquantifiable.

Evaluating a US federal courthouse

Since 2003, the US General Services Administration (GSA) has developed a National 3D-4D BIM Program. The GSA is the largest real-estate organization in the United States, providing workplaces for 1 million federal workers. (Refer to Chapter 9 for more information.) Continuing with the law-and-order theme from the sidebar where we discuss the Cookham Wood example, GSA selected the US Federal Courthouse in Jackson, Mississippi, to be a GSA BIM pilot project.

The project team identified more than 7,000 interferences and clashes during the project and resolved them during design. The client also achieved savings through reality data capture, such as laser scanning. In a test to demonstrate the efficiencies and savings that BIM brought to the project, the project team compared material schedules extracted from the digital model to material quantities that are produced with traditional estimating methods. The comparison showed that information taken from the digital model was more accurate.

Calculating the Savings

Industry reports suggest that investment in BIM implementation is paying off, but as yet there is no formal common agreement as to how savings and return on investment for BIM are calculated. That isn't to say, however, that there are many benefits to be had. In the following sections we explain what benefits you can gain immediately and how these benefits are calculated from the logic of BIM.

A number of reports and research have indicated that the construction industry is seeing a positive return on investment. In a report called "The Business Value of BIM for Construction in Major Global Markets," undertaken by Dodge Data & Analytics (formally McGraw Hill), findings showed that two-thirds of BIM users in the United States said that their investments were paying off. Alongside anecdotal evidence, Dodge Data & Analytics has formally measured case studies and early adopter trial projects to record both the investment and the return in commercial terms.

Coming to a formally agreed way in which you can measure benefits such as cost is difficult. No standard metric exists for measuring the return on investment (ROI) on BIM. A report by buildingSMART acknowledges that it's hard to put a price on work that was either avoided or extra work that didn't take place due to a BIM process and therefore difficult to calculate a true ROI for BIM. buildingSMART is keen to encourage the uptake of a standard methodology for measuring performance. The measures that it puts forward as items to consider for a uniform measurement process include the cost of preparing a facility management (FM) database, construction tender value, outcome profit, and a number of soft indicators, such as quality of information and the elimination of pain points where problems often occur.

The National Institution of Building Sciences (NIBS) reviewed the performance of projects and indicated that the construction industry can achieve savings (up to 4 percent on construction of a new build and 1.5 percent on refurbishment) by using BIM for information management. Unfortunately, as with many studies, it didn't explore savings due to BIM in the operational and maintenance stages. The BSI Investment Report paints a similar picture. Most of the data covers design and construction, but the report indicates that when you consider that 80 percent of an asset's cost is during its operational phase, BIM has considerable benefits post-construction still to measure.

Plenty of benefits are to be had through BIM across the entire project lifecycle. In the next section we explore some of the immediate impacts that BIM could have on your organization.

Measuring the immediate impact

Although you reap many benefits in the longer term, plenty of benefits make an immediate impact, including

- **More for less:** BIM provides better value for money. Another way to look at cost savings is to think of the outputs rather than individual examples. Instead of showing that BIM can save you, say, 20 percent on a project, demonstrate that you can produce five buildings for the price of four.

- **Right first time:** Because you and the project team are all using and sharing the same base information, there are more opportunities to use up-to-date information, meaning that you get it right first time.

- **Build better before built:** Building digitally first before committing to the actual building of the physical asset means that you can test and analyze the data first, resulting in an optimized design before construction begins.

- **More innovative solutions:** Through BIM you can explore many different opportunities in a relatively short space of time and test the outcome and performance of these options, resulting in innovative solutions.

- **Better collaboration:** The BIM process brings the project team together through common standards, better communication, and the use of a common data environment, all of which contribute to better collaboration.

- **Retention and attraction of the best staff:** Adopting lean processes and procedures puts you and your organization at the vanguard in new developments and ideas. What member of staff wouldn't want to be part of that?

Expecting savings based on the logic of BIM

BIM achieves the real savings through coordinated information, reducing the need for rework, increasing efficiency, and improving information accuracy by working from the same base information. If you're a client buying your data for the first time, remember that the cost savings occur throughout the whole project timeline:

- **Soft landings:** They seek to reduce the hidden costs of adapting completed spaces to suit specific end-user needs.

- **Increased use of standardization:** It generates efficiency and procurement savings such as through the use of offsite manufacturing.

✓ **Better team collaboration:** Contractors, suppliers, and consultants work harder to beat benchmark targets, leading to better competitiveness, which also leads to less risk, lower insurance premiums, and a reduction in potential claims.

✓ **Visualizations and 3D models:** They allow the clients, project teams, and stakeholders to understand the design intent, explore design options, optimize deliverables, and reduce client variations.

✓ **Standard documentation:** Employer's information requirement (EIRS), BIM execution plans (BEPS), standards, and protocols allow the project team to coordinate and share information and documentation in a timely and consistent manner. They agree what standards the project team will use and what the deliverables are.

✓ **Single source of truth:** Working within a common data environment (CDE) allows all project members to work on the same base information, reducing the need for rework. The BIM process can detect clashes and conflicts early and resolve them, leading to a reduction in waste and delays.

The potential benefits of BIM are only going to increase as processes, technology, and adoption improve. Level 2 BIM has limited functionality for delivery of operational data sets and integration of telemetry. However, as you move into the future, Level 3 will address these elements in sector delivery and operational stages, with a focus on *enabling total cost,* which is the amount of money spent by an organization to produce a given output and carbon outputs. Operational data will help unlock savings in asset performance.

Separating Reality from the BIM Wash

Pilot projects should be well evidenced with a clear breakdown as to how the client realized savings. A little white lie here, a slight over-exaggeration there — everyone gets carried away from time to time. The term *BIM wash* describes overinflated claims of using BIM process, products, and services.

Sometimes this overexaggeration is innocent and unintentional. Other times people don't fully understand BIM. (Chapter 23 discusses some common myths that people may believe about BIM.)

More worrying are the individuals and organizations that engage in BIM wash intentionally, from individuals offering consultancy services that don't live up to clients' expectations to software companies making bold, unsubstantiated claims about their "total BIM solution." To avoid this confusion in the industry, state what your end deliverable will be. Being able to calculate the benefits of BIM based upon tangible metrics is one way in which you can reinforce how BIM can improve process, products, and services.

In the next section, you consider how BIM relates to other construction disciplines.

Evaluating BIM for all construction disciplines

BIM isn't just about the building or asset envelope but also how that asset connects to the landscape. Infrastructure plays a vital part of construction, yet in many instances you may not see those projects because they're buried deep beneath the surface. Infrastructure projects such as tunnels, railways, and roads are often of utmost national importance and are important to a country's economy.

As Chapter 2 explains, the term Building Information Modeling is a bit misleading, because BIM is a verb. Instead of thinking about building as a structure with a roof and walls or an envelope, think about the term *building* as in the action of constructing or building something. BIM isn't just about design. It covers the whole project timeline, including post practical completion and project management.

Putting BIM into practice during the operation phases

The *operation phase* of an asset usually commences as soon as the owner occupier takes possession at the handover stage after the project completes. This is the phase when the client or end user starts to use the built product (or service). Asset and facilities management considers optimizing whole-life costs. This could be over a portfolio of assets, all of which may be completely different to each other and even miles apart.

The digital information collated during the design and construction is vital in understanding the asset's future operational and maintenance requirements. It allows the client to make better informed decisions and conduct better scenario planning on operation and maintenance costs because it's based on actual statistics and facts on the asset's performance and status. There is also an increased amount of verification of information because information is based upon quality information. The client can also make better organizational and strategic planning decisions because asset information is more accurate and complete, and the client can more easily generate legislative documents such as health and safety files or operational and maintenance manuals.

The BIM process can also reduce costs by transferring accurate, consistent, clear, and complete information, at handover stage or when transferring operations from one service provider to another service provider.

BIM can greatly improve the handover process for clients by delivering information that's both structured and consistent, such as the specified information set COBie. This approach to information delivery removes any complexity from process, technology change, and competition issues within the supply chain. (Refer to Chapter 11 for more information about COBie.)

Chapter 16

Using BIM On-Site and BIM for Health and Safety

*A*lthough all the parties engaged in the construction process have a contract, this chapter specifically talks about the main, general, prime, or principal contractor, as in the person or organization who undertakes to perform the services, supply goods, or carry out work necessary for the construction of a project. The concept of a project team comprising the client, design team, and contractor is still valid and relevant in this BIM era. However, the industry is witnessing a cultural change, with better-informed clients, a range of procurement routes, and early contractor engagement through initiatives such as soft landings.

By involving the contractor early in the project, the whole project team can consider buildability issues and construction planning earlier in the design process, benefiting from the contractor's and supply chain's knowledge and experience and leading to innovative solutions.

At the most simplistic view, the project team consists of the client, design team, and contractor. Early contractor engagement, better informed clients, and new procurement routes like design and build mean that there is no longer one standard approach. A traditional structure is sometimes used for the early stages of a project and then changed to a contractor-led project team, which is where the design team's contractual rights and obligations are transferred to the contractor, which is sometimes called *novation*. The two project structures can be described as:

- ✔ **Traditional project team:** The client appoints the design team to produce a design and develop certain details. A number of contractors tender for the project and the successful contractor builds the project.

- ✔ **Contractor-led project team:** The contractor leads the project team with the design team forming part of the contractor's team. A number of contractors bid for a project, based on a comprehensive brief, and the client decides on the winning tender on the basis of design, cost, and other factors.

In this chapter, you discover how improving the information at the front-end of BIM through early contactor engagement makes for savings during the construction process, and how the project team is using BIM to produce record information, such as as-built models, operation and maintenance manuals, and health and safety files that are accurate and updatable. Furthermore, this chapter looks at how BIM can make your assets safer. After all, the loss of life is almost unquantifiable.

Examining Contractor-Led Project Teams: Planning for Quality

Contractors come in all shapes and sizes, and their services are wide ranging. Some contractors perform work with the help of specialized subcontractors, whereas others just manage the work and hire subcontractors. Some contractors offer services beyond the handover phase, and as well as building the asset they then operate and maintain it. Then you have contractors that also act as developers; a prime example is house builders.

These sections examine the many benefits that BIM offers contractors and how an information-rich model can be used on-site in a number of ways. In order to make the most of BIM for the contractor, you need to know how the contractor uses your information in order to provide exactly what the contractor requires.

Eyeing the contractor's benefits of utilizing BIM

Contactors use BIM to virtually design the asset first before a spade even goes into the ground. This brings with it new opportunities for innovation, resulting in a better-quality end deliverable. The contractor who uses BIM experiences the following benefits:

- **Improved tendering:** Getting the contactor's and supply chain's early engagement leads to a better end product, not only in terms of quality, but also in terms of cost understanding and making construction programs shorter.

- **Visual communication:** The BIM process enables visualization of construction products and processes, which helps all the stakeholders in a project. The project team shows clients and stakeholders visually how the asset is best constructed and has the opportunity to engage non-technical people, which can help clients to make decisions.

- **Site organization and planning of temporary work, workforce, and traffic management:** As part of enabling construction to begin, the contractor will be required to plan out just how the construction site will run and operate. The use of BIM, in particular linked to the time element known as 4D, enables the contractor to plan out in what order construction will commence.

- **Clash detection:** The ability to identify collisions and mitigate them before they're constructed is a result of the whole team using the same base information and sharing and communicating with each other. This in turn results in fewer clashes on-site.

- **Reduced number of design information requests:** Coordinated information results in a better understanding of the end deliverables and fewer errors because everyone has an increased chance of using the most up-to-date information.

- **Off-site fabrication:** This refers to higher reliability of expected field conditions, allowing the opportunity to do more prefabrication of materials off-site, which usually means a higher quality at a lower cost.

- **Scenario planning:** The addition of time and scheduling allows the project team to virtually construct the asset before the contractor builds it. The contractor can explore a number of what if scenarios, such as looking at various sequencing options, site logistics, hoisting alternatives, and cost.

- **Improved health and safety:** The BIM process brings with it opportunities for integration of temporary works into the design. This in turn leads to increasing safety-managing public interfaces.

- **Verification:** Receiving information in a consistent and structured way results in data that can be validated and checked. Verifying data and information is much easier for the contractor if she receives it in a digital form, particularly if it's an open data format such as IFC, which many upstream collaboration and checking tools use.

Automating the code checking process

In order to adhere to existing procedures for building approval, the comprehensive 3D information models provided by BIM are often printed out into a bunch of 2D plans, elevations, and sections on big sheets of paper for submission, which is often frustrating for both design teams and code officials. Project teams are benefiting from the efficiency benefits of BIM, but this hasn't really extended to code checking officials.

Recently, the US International Code Council and a nonprofit consortium of international technology companies, such as Solibri and Avolve, called Fiatech and based in Austin, Texas, demonstrated a proposal called the AutoCodes project. The aim of AutoCodes is to transform the traditional process of reviewing building drawings and automate code checking using BIM. By checking models against the pre-installed code rulesets, you can visually appreciate any changes to your project and make the required alterations or discuss them with colleagues. You can find out more about the project and the development of an open-source ruleset library at www.fiatech.org/the-autocodes-project.

Take a look at Chapter 20 to find out how a Google spin-off company, Flux, is also looking to make it easier to check building code requirements there with "Flux Metro."

Additionally, in Singapore, the Construction and Real Estate Network (CORENET) e-submission system is an Internet-based system that enables industry professionals to submit project-related electronic plans and documents to regulatory authorities for approval and then track the submission status online. The system handles project-related documents for the whole project lifecycle, covering processing of plans and documents related to issuance of

- Planning approvals
- Building plans approvals
- Structural plans approvals
- Temporary occupation permits
- Fire safety certificates
- Certificates of statutory completion

You can now submit approvals earlier at a planning stage, and with BIM the industry is moving to approval rather than supervision. After the project gets to site, the use of surveys, augmented reality, and tagging ensures that what you have in the digital model translates onto site.

Using site information

An information-rich model is full of useful information that the contractor can use on-site in a number of ways and for a number of purposes. The preparation of the model during the tender process enables the contractor to understand and challenge the designer's details, bringing together the contractor and supply chain experience and input. The contractor can reuse and present graphically, visually, and sequentially information that the architects and engineers have historically prepared during the tender process.

The contractor can develop the project information model (PIM) to incorporate a site setup model. The PIM can show the locations of welfare facilities, such as site cabins and toilet facilities, as well as laydown areas (for temporary setting down of materials) and storage areas for construction products. By integrating the PIM to the element of time, 4D sequencing can be achieved, which is useful in demonstrating to the client and project team how the contractor intends to construct the project.

The next sections talk about a few of the different ways in which the information-rich model can be used and the best way of viewing the information when out on site.

Utilizing clash detection

The contractor coordinates information from subcontractors and other consultants and as such looks for clashes. Historically, you carried out *clash detection* by overlaying 2D drawings, but with BIM you can look at clashes in the third and also fourth dimensions. *Hard clashes* are when something occupies the same space or penetrates something else. Also consider clashes in terms of clearance or operation space; for example, make sure components have enough clearance space around them so they can be maintained or that a door swing has adequate clearance when open.

As you discover in Chapter 21, often the project team carries out clash detection via collaboration and construction management tools or standalone model-checking software. You also need to consider using information to best avoid clashes with regards to deliveries of site materials. For instance, you don't want a truckload of insulation turning up to site at the same time that the school across the road is finishing for the day. Trucks and children don't mix.

Viewing site information

The way in which you distribute data and information is changing to digital, and therefore you need to consider who's going to use the information and how they're going to view it. For the contractor-led team that spends more time in the field working on the construction site, access to a desktop computer in a cozy office may not be a feasible option. Other methods such as mobile devices and robust workstations are more appropriate.

- ✔ **Computer tablets and smartphones:** Traditionally, you kept paper-based drawings on-site and perhaps even laminated some to protect them from the elements. As the industry moves toward a paperless environment, computer tablets are becoming commonplace on the construction site, to both receive and review information.

  Review your company policies with regards to mobile tablets and smartphones, because many organizations prohibit their use on-site.

- ✔ **Site workstations:** A *site workstation* is simply an access point at which site operatives can access project information. Think of an office desk, computer, and chair, but stuck in the middle of a building site. Contractor Balfour Beatty has developed a SmartBox as a way of sharing project information on-site. The SmartBox is essentially a computer, printer, screen, keyboard, and mouse, housed in a large metal box that the contactor can move around the site with a forklift truck.

- ✔ **LCD screens:** The price of flat-screen TVs has fallen considerably over recent years, and including one in your site welfare cabin or site office is a simple and effective way to display project information to the site team. Communicating in a visual way is a great advantage to highlight relevant information to construction workers. This is especially important because site operatives often change over time, coming onto site for a particular role or task and then moving on to the next site, meaning that training may need to be repeated several times. Also, communicating visually can cut through any language barriers in situations where not everyone speaks a common universal language.

Answering contractors' questions

The contactor focuses on the people who have a stake in the project, including generators, reviewers, or receivers of information. (Refer to Chapter 3 for more information about these different people.) The contractor is also probably receiving model files in a number of different file formats, and so a model integrator is a good way in which to bring all this information together so that the project team can review and analyze it.

Historically, tasks such as cost estimating, coordination, and scheduling were time consuming, error prone, and costly. Why? Because the contractor used 2D plans and paper-based specification documents to manually perform quantity take-offs, often creating a model from 2D paper-based drawings to service needs. Add into the mix that the contractor was historically not engaged early in the process and therefore unable to impart critical knowledge and wisdom, and it's little wonder that these tasks were often completed very late in the day.

With the advent of BIM, new methods of procurement, and soft landings, project teams are now providing models earlier in the procurement process, meaning that the contractor can use this base information for tasks such as estimating, coordination, and fabrication. This often involves further development of the initial model by the contractor and additional detailed information that's not traditionally provided by the architect or engineer, such as method statements or production rates. As a minimum, the contractor needs a model that contains the following information:

- ✔ **Detailed construction information:** The model should provide accurate and consistent 3D views, sectional information, elevations, plans, details, and schedules so that the contractor can extract the relevant information for quantity take-off. Many contracts still rely on 2D information as a contract deliverable. However, having a 3D model enables the contractor to extract, use, and manipulate the information in a number of different ways for multiple purposes.

- ✔ **Specification information:** The specification should cover the performance, execution, and maintenance of products. Using the specification, the contractor can make any product selections or product purchases, and analyze any data related to performance requirements and project deliverables, such as acoustic levels or fire resistance. Ideally, the specification should link to the geometric model information for coordination purposes.

- ✔ **Temporary components:** Information and objects representing equipment, formwork, and other temporary works structures help the contractor when planning any construction sequencing and with site planning.

- ✔ **Status of information:** All information has a purpose and should also have a status to identify the suitability of the information provided and for what purpose it can be relied upon. For example, at an early stage the model may only be developed enough to show spatial coordination, whereas at a later stage the model is developed so that contractors can use the information for contraction purposes — that is, to actually build the asset using the information provided.

You may need a variety of software applications to produce the preceding list, because no one BIM platform can deliver all the required deliverables. Therefore, consider tools and applications that offer interoperability.

Connecting the feedback loop

Today's clients aren't just procuring buildings; they're also obtaining digital information. The project team adds, embeds, or links information to the model during the construction, so that the client, owner, or facilities manager can extract the information at a later date.

Unfortunately, one drawback to being human is that errors do occur. Although the digital model may be accurate, the contractor has to verify that what has been constructed and installed meets the project specification in terms of its performance requirements and end deliverables. The digital virtual building model acts as a reference point to check against any field errors.

In the following sections you discover that verifying the as-built information involves a combination of traditional methods, such as daily site walks and inspections through to automation verification techniques using digital technologies. For example, at the design stage, you can buy software to check that a building's design complies with the Americans with Disabilities Act (ADA) standards for accessibility. The future of BIM is allowing you to verify that what is constructed on-site also matches those requirements.

These technologies can be categorized as follows.

Tagging

Tagging has been commonplace within the retail industry for many years as a way to track and analyze stock levels. By using a code or series of numbers, items can be tracked back to their source or origin. Perhaps you bought this copy of *BIM For Dummies* from an online retailer. (The authors thank you, by the way, for a sound investment.) From the moment you entered your credit card details, this copy of this book was assigned a job number and then tracked right through from the warehouse to the sorting office to the moment it was delivered to you.

Contractors are using radio-frequency identification (RFIDs), quick response codes (QR), and barcodes to track building components by transmitting information about the tagged object back to a database. Contractors can then view information about delivery, installation, and commissioning via a mobile device. Unlike barcodes, however, RFIDs allow you to identify tagged objects wirelessly without requiring a scanner to physically scan the item. As the cost comes down and the technology matures, in the future components such as individual bricks could be tagged, which would allow for accurate inventories and eliminate waste.

As well as tracking building components, tagging can also capture as-built information. The virtual component in the model is associated with the actual installed component, which helps you to locate it if it requires maintenance in the future. This information is highly valuable to facilities management.

Mobile computing and field BIM

Today's cell phones are more portable digital media hubs than devices for making and receiving calls. They take pictures, they act as a satellite-navigation, they play and store your digital music, and they allow you to send emails. Some say that today's average mobile phone has more computing power than *Apollo 11,* which put a man on the moon. Mobile phones have revolutionized the way in which people communicate with one another, and they give the user instant access to information. Aside from texting, tweeting, and taking the occasional selfie, you can use mobile devices as great assets on the construction site. With information increasingly stored in the cloud, mobile devices give you the ability to collect, store, retrieve, review, share, and update information — in some cases, directly with the model.

Tablet computers, smartphones, and glasses linked to GPS technologies and the Internet allow contractors to walk around the site prior to any works commencing and see via the device the works that are to be constructed in that area of the site. *Augmented reality* uses a combination of viewing the physical real world in real time and a digital overlay or augmented model information over the top. Imagine standing in a field and being able to see where the future building, road, or tunnel will eventually be, or looking at the ground and seeing a digital representation of where existing service pipes and sewers are located.

Add to the equation object recognition and the experience becomes fully immersive and interactive. This not only allows you to see where construction will be carried out but also to review any areas of difficulty.

Laser scanning

One technology and area of the industry that has had a rapid growth expansion through technology is *laser scanning,* which is basically the use of controlled steering of laser beams followed by a distance measurement that is taken at set increments at every pointing direction. Laser scanning can quickly capture shapes of objects and is not only used for buildings but also for capturing the contours of the landscape.

Although it has been around for a while, the construction industry has only recently felt scanning's adoption within recent years. Existing buildings and retrofit, ancient, and historical assets are all benefiting from this technology. The value of scanning also extends to capturing data in hard-to-reach places and where existing documentation is lacking or even nonexistent.

3D scanning is more than just a 2.0 version of traditional surveying techniques. Yes, it has improved surveying time and accuracy of data, but it provides many more opportunities besides, including the ability to monitor, assess, and analyze physical data. As with all technology, the cost of laser scanning equipment will go down and the quality and scanning speeds will increase over time, making it even more commonplace. A 3D scan can create the foundation for a BIM approach because you're starting off with good-quality 3D base information to work with.

3D scanning, compared to traditional surveying techniques, is generally more expensive when you take into consideration the cost of equipment, mobilization, and process times. That said, the benefits and opportunities that it can bring mean you should consider return on investment carefully. In some cases traditional surveying techniques may not be possible, particularly on complex or challenging projects.

Here are two guides to scanning that we suggest you read:

✔ The GSA's Building Information Modeling Series 03-3D Laser Scanning guide (in draft) provides information on 3D imaging. Check out www.gsa.gov/portal/content/102282 for more information.

✔ The UK BIM Taskforce (www.bimtaskgroup.org/wp-content/uploads/2013/07/Client-Guide-to-3D-Scanning-and-Data-Capture.pdf) has produced a handy free-to-download guide for clients, explaining what they need to consider when procuring 3D scanning.

Requiring BIM Use On-Site

BIM isn't just about drawing models in the office or improving design processes. Some of the biggest benefits of BIM implementation take place on-site. In the following sections, you can see the impact of BIM on site safety, encouraging greater site efficiency, and developing skills. BIM is often associated with three-dimensional (3D) interpretation of built assets, but by combining that data with an understanding of how it changes over time in what is called four-dimensional (4D) BIM and how it affects cost in five-dimensional (5D) BIM, you can develop a much more in-depth understanding of your project than ever before.

Promoting a modern construction industry

The construction industry undoubtedly requires a big change to shake off its present-day image. Change is required in the construction industry itself and in how the public perceives the construction industry. Industry and government must work together to inspire young people. Currently, the industry has a lack of career attraction with young people preferring to go into other industries that they perceive as more rewarding. This is due to perceived poor image, lack of gender diversity, low pay, and lack of job security due to the cyclical nature of demand for construction. All this is especially evident in construction contracting.

In order for the construction industry to be the career choice of young, talented students coming out of school, colleges, and universities, the industry has to pull up its socks and up its game. The industry can do so by improving quality, addressing the skills shortage, and removing barriers to innovation, all of which BIM can help with.

Improving quality

BIM is helping to improve the likelihood of quality construction through early contractor involvement and also safer construction, which we explore in more detail further on in this chapter in the "Reducing Risks On-site: BIM for Health and Safety" section.

The Considerate Contractors Scheme (www.ccscheme.org.uk) aims to improve the image and quality of the construction industry and of contactors in particular. It's a nonprofit-making, independent organization in the UK that invites construction sites, companies, and suppliers to voluntarily register with the scheme and agree to abide by the Code of Considerate Practice, which is designed to encourage best practice beyond just statutory requirements. The scheme is concerned with any area of construction activity that may have a direct or indirect impact on the image of the industry as a whole. The main areas of concern fall into three categories: the general public, the workforce, and the environment. Registered sites and companies also have to address how they'll encourage new people into the industry with regard to careers advice, apprenticeships, and placements.

The Construction Skills Certification Scheme (CSCS; www.cscs.uk.com) provides proof, by way of a card, that individuals working on construction sites have the required training and qualifications for the type of work they carry out. The scheme keeps a database of people working in construction who have achieved, or are committed to achieving, a recognized construction-related qualification.

Addressing the skills shortage

Industry and governments need to work with skills providers to assess the industry's future needs. There need to be clear entry points at which the next construction professionals can enter the profession; for example, clear educational routes, courses, and apprenticeships that will lead to a rewarding career within construction at the end of it. The industry needs to explore making apprenticeships more flexible, making training programs that fit both around the employer and student, and reviewing current skills delivery mechanisms. The UK Contractors Group (UKCG) represents leading contractors operating in the UK and it's provided a great resource called Born to Build (www.borntobuild.org.uk). The site aims to answer questions from the next potential generation of construction professionals, such as "How can I get experience?" and "What qualifications do I need?". It also dispels myths such as the industry is only for men.

Removing barriers to innovation

Analysis shows that around two-thirds of construction contracting companies aren't innovating at all. Alternative collaborative procurement models are replacing adversarial culture and are demanding both cost reduction and innovation within the supply chain as a way to maintain market position. This replaces the old school of thought and approach that focuses innovation on the bidding process with a view to establishing a bargaining position for the future.

Contractors are using BIM and virtual construction to come up with new methods, new innovative ways of building, and safer techniques to set themselves apart from other contractors.

Ensuring accuracy and removing waste and inefficiency

The approach to risk across the supply chain drives much of the waste in construction. The supply chain procurement processes can be bureaucratic and wasteful. A landmark report called "Rethinking Construction" in 1998 by Sir John Egan, a former chief executive of Jaguar, highlighted a number of inefficiencies in the UK construction industry. The report stated that at least 10 percent of materials and 40 percent of manpower on construction projects are wasted, and 30 percent of construction is rework.

To promote a modern industry that the construction sector deserves requires changes to processes and its approach to constructing built assets. The following sections discuss a few initiatives that the construction industry can take to ensure that it is a lean, mean, accurate, and efficient industry, all of which go a long way to promote a modern construction industry that everyone can be proud of.

Getting the contractor involved early

Acquiring the right knowledge and expertise early on can pay dividends in the end. Early engagement of the contractor means that you can tap into all of that experience and ensure accuracy of the produced information. It's not about getting the contractor in and changing the whole design. It's about working together to get the best solution for the client.

Change is inevitable during the design and is part of the process as you look for the best solution to solve a problem. The problem is when that change comes late in the processes — whether due to late changes by the client, unforeseen circumstances, or because of mistakes — as the ability to impact change decreases over time, while the cost to make changes goes up. Working early with the contractor can help minimize this and avoid costly waste. This has been highlighted by Patrick MacLeamy, chairman of HOK and founder of buildingSMART International, in his concept of the *effort curve*. This has become known among BIM advocates as the MacLeamy Curve. However, it draws on work in the 1970s by Boyd Paulson, a Stanford civil engineering professor. His original diagram indicated how the level of influence on project costs decreased over time.

Patrick MacLeamy's interpretation of Paulson's concept shows that BIM requires an earlier engagement of project stakeholders so that all can coordinate the design, encouraging a more integrated approach to project design and delivery. He advocates that resource associated with design should be moved forward in the project timeline to earlier stages. To eliminate construction rework, early engagement of the contractor and the supply chain is vital.

Moving beyond throwing things away

With BIM technology, integrated design and specification processes can remove waste in materials and misdirected activities. All construction work has tolerances that allow for inherent variances in construction materials and workmanship skills and are tolerated if commonly accepted construction industry standards are adhered to. While trimming off the odd inch here or cutting off a few feet there on-site may seem insignificant, added up over the whole asset that's a lot of waste. With BIM, tolerance will become a thing of the past, because you can so accurately plan something like plasterboard sizes or precast panel construction, so you don't need to specify tolerance. Inefficient and dirty/messy methods, like concrete pouring on-site or hacking bits of steel to fit, will disappear because you know exactly how much material you require.

This in itself will stop overproduction of materials and also reduce the amount of materials that you hold on-site. Less material also means lower transportation costs. Be resourceful! Resource-efficient construction makes the best use of materials, water, and energy over the lifecycle of built assets to minimize embodied and operational carbon. Consider reused materials or materials with recycled content. The WRAP website (www.wrap.org.uk/resource-efficient-built-environment) has a number of free tools to get you going.

Waste isn't just a site issue. Decisions the project team makes during the design stage can greatly affect the amount and type of waste created, and the ability for the contractor to reduce, collect, and recycle it. The whole project team, including the client, has a responsibility.

Making the most of BIM off-site

As the demand for low carbon and sustainable construction increases, the potential for off-site construction techniques to deliver faster assets with less waste and energy will become ever more popular. Off-site construction involves building parts of the asset away from the construction site, usually in a dedicated factory, and then these are delivered to the construction site for assembly (if not carried out in the factory) and installation. Compared to traditional build techniques, off-site construction has the potential to greatly minimize on-site waste. Unlike the construction site that is open to the elements, production is carried out in an environment-controlled factory where it is easier to build to higher standards. This process does mean that some waste transfers to the factory environment; however, the amount is significantly less than building on-site. The benefits of off-site construction include

- Greater precision
- Increased quality
- Reduced overall manufacture/assembly time
- Safer and cleaner working conditions

BIM is enabling contractors to feed in directly to this process as you can read about in the "Mobile computing and field BIM" section earlier in this chapter.

In 2013, the National Institute of Building Sciences established the Off-Site Construction Council (OSCC). It serves as a research, education, and outreach center for relevant and current information on off-site design and construction for commercial, institutional, and multi-family facilities.

Timing the construction process (4D)

You're probably familiar with the concepts of 2D and 3D, but here we introduce you to another dimension, 4D. In the past, a disconnect existed between CAD and project management software, but now you can benefit from linking the geometry to a timeline or schedule. As you find out in Chapter 21, construction management tools allow you to take an overview or to holistically review model information. By integrating the 3D model with popular project schedule applications, you get a 4D model.

Adding the element of time gives you some interesting possibilities because you can plan for construction phasing over time and build the asset virtually; think of it as a dress rehearsal before the big event. Construction sequencing, simulations, and scheduling capabilities allow you to highlight and mitigate any risks within the sequence.

The New York City Department of Buildings has a number of site-based BIM objects for free download at `www.nyc.gov/html/dob/html/development/bim.shtml`. They include objects such as site cabins and cranes, which can be beneficial in site planning and setup.

Estimating made easy — quantities and costs (5D)

The concept of 5D focuses on adding the element of cost. Today the concept of counting brick courses or numbers of windows seems just ludicrous, but the construction industry was doing this only a few years ago. Although no magic button exists to generate cost information, BIM does make this a lot easier. It allows contractors to extract material quantities from the model. Doing so enables the contractor to accurately measure and quantify the elements of the structure efficiently, and to a greater degree of detail than was previously available. The provision of detailed quantity schedules for materials enables contractors to

- ✔ Remove risk from their pricing models.

- ✔ Ensure that clients are being provided with quotations that are based on precise data that is consistent across all the parties tendering for a contract.

Historically, the lowest tender price to build the job has usually been the most favored. With the accuracy and certainty of tender prices increasing, contractor competition will be based on quality as well as long-term guarantees of performance and low maintenance, instead of just being just the cheapest. Accurately assessing cost and quantity information enables contractors to not only buy more efficiently but also consider just-in-time procurement and delivery of products. Buying products not only at the right price but also at the right time when you require them is important. This limits costly on-site storage of materials and limits the opportunities for those not-so-honest people who sometimes like to help themselves to building products and materials that aren't theirs!

Cost estimations will increase with accuracy as the model develops over time. At the design stage, cost estimating is carried out on the available information, which is likely to be on areas, volumes, and space types. Then the project team has to develop the model enough for parametric cost estimation, where costs are based on major building parameters such as number of floors and roof areas.

In order to get the best results, the project team has to define objects and components so that other people in the team can extract information. For example, a stud partition wall is just a rectangular geometric shape unless it's been defined as a stud partition wall. BIM software can give you the length, width, and construction of the wall, but it may not be able to report on every stud. Although BIM can greatly help with cost estimation, consider it as a wider holistic approach. Quantity surveyors may export model information for use in a dedicated estimating software package, or in some cases quantity surveyors link information to estimating software with add-ons and plugins, and review information in a spreadsheet format.

Giving fabricators what they need

Just as the design process benefits from the contractor's early engagement, early engagement of the subcontractors and fabricators can reduce errors and speed up the processes. Fabricators act as a hybrid of manufacturer and contractor. Some fabricators, such as steel fabricators, produce a range of standard, off-the-shelf products as well as custom-made products for a particular project requirement.

In other industries, such as automotive, designers, and engineers export CAD model information to the production processes, where it's used in computer-aided manufacturing (CAM) and computer numerical control (CNC). Construction products such as doors and windows are produced in this way, but require information additional to that typically produced within BIM.

Digital modeling can come very close to fabricating modeling, but in most cases the project team uses BIM to imply a product rather than fabricate it; for instance, the designer provides enough information about a product so that the contractor can go away and purchase it.

Off-site fabrication can bring about a number of benefits for the contractor. Through the use of BIM, the contractor can feed information into the process, such as 3D geometry, material specifications, any finishing requirements, and when and in what sequence the items will be delivered to site.

Reducing Risks On-Site: BIM for Health and Safety

Unfortunately, despite the best efforts of all concerned, too many people are killed and injured in the construction industry each year. Even in times of economic downturn, construction is still one of the largest industry sectors and also one of the most dangerous. Although the commercial advantages of BIM are fairly obvious, the potential benefits in improved health and safety, by reducing accidents and deaths, are so great as to be almost unquantifiable.

The potential to both influence and prevent construction injuries decreases exponentially as the project progresses. Therefore, the most effective safety program elements occur at the planning and pre-construction phases of a project.

The contractor has a duty to plan, manage, and monitor an asset during its construction. This involves liaising with the project team, including the supply chain, to reduce risks on-site.

Innovations such as off-site manufacturing reduce waste on-site and also go some way toward a reduction in accidents and injuries on-site because components are constructed in a safe and controlled factory environment. However, when it comes to the installation process on-site, each location is unique and you can't always predict conditions. Plus, the process of delivering large pre-assembled components to site brings its own health and safety challenges.

Design for Safety (DfS) is the process of *designing out,* which is to mitigate the hazard during the design phase or minimizing health and safety hazards at the earliest instance. It's a similar concept to the US initiative Prevention through Design (PtD) in which you mitigate occupational hazards during the design phase. Both these concepts work well within the ethos of BIM in that they promote collaboration between the project team throughout the whole delivery process.

Often poor integration exists between construction processes and health and safety issues, further compounded by a lack of proper job-specific health and safety training. Although contractors comply with legislation and paperwork, the most effective form of communication transfer is visual. Historically, the construction industry has communicated with 2D drawings. But with BIM, as you move from a document to a database system, you can extract assessments of risks and hazards from the model in a report form and then specifically target the person who's best placed to manage them. The following are simple but powerful tools that can be used to reduce risks on-site:

- **Tracking and sensing:** A commonly cited benefit of BIM is the ability to detect clashes. Examples are usually shown of a service pipe clashing with another service pipe, all of which are beautifully color-coded. However, in a site context you have to consider physical clashes. For example, many accidents and injuries occur on construction sites across the world when workers are struck by moving vehicles, equipment, and machinery. Technologies such as small proximity sensors are becoming more commonplace. You wear the sensor discreetly in a hard hat, and it gives an audible alarm or vibration if you're too close to a crane or plant machinery. In this way you can establish no-go areas on-site.

- **Simulation:** As we discuss in the earlier section "Timing the construction process (4D)," through the additional element of time you can begin to rehearse the construction sequence. 4D is particularly useful for site safety planning, enabling the project team to visualize safety arrangements at different moments in time during the construction process. Image you're working on a new-build hospital project. Knowing the best lifting arrangements for craning in a large piece of medical equipment is very beneficial.

- **Worker safety training:** BIM is a powerful communication tool that can aid understanding among members of the workforce. Virtual reality and augmented reality can aid understanding before a site operative carries out a task, and a simple LCD screen in a site welfare cabin can show accident blackspots and display model information, giving all site operatives access to coordinated information.

Chapter 17

Identifying All the Users of BIM

In This Chapter

▷ Coordinating what the BIM team members require

▷ Assigning BIM roles and responsibilities

▷ Working out the people you need to help generate and manage BIM content

▷ Making the most of BIM for asset and facility management

*U*nderstanding that the same BIM and the outputs it can generate are going to be used by different (and new) roles in the industry, at different times and in very different ways, is vital to seeing the point of BIM. This chapter looks at BIM users and roles from inception to demolition (and beyond). We discuss facility management (FM) users and operation toward the end of the chapter.

Anticipating the Information You Need

The project timeline has multiple users. Don't just think about your own segment or silo but also how the information is going to be used by everyone on the team and the built asset in the future. The following sections examine who needs what information across the project and for what purpose as well as the different information needs of other members of the project team, the timing of that information, and when you need to deliver it.

Eyeing information across the project timeline

Professional institutions have restructured plans of work to take into consideration the early work flows required for the BIM process and to encompass the whole project lifecycle, from identifying strategic needs through to operations and end of life. Furthermore, they organize the process of briefing, designing, constructing, maintaining, operating, and using construction projects into a number of key stages. The amount and level of information increases as you progress across the project lifecycle. (Refer to Chapter 9 for more information about the timeline.)

In some instances one party can't progress because it's waiting for information from another. For example, the structural engineer may be waiting for the architect's general arrangement drawings before progressing further with the structural design. BIM tries to solve this by putting information into a central depository; however, the process still relies on all parties to pull their weight and exchange information in an agreed and timely manner.

Here we discuss the different project team members' needs and the types of information they require.

Clients' needs

Clients are going to use and expect different pieces of information from the model. Initially, this information comes chiefly from the design team. Clients are looking for outputs; for example, the as-built model or visualizations, which the client or client's agent may use for marketing purposes. They're not interested in details such as structural steel layouts or drainage runs, but are interested in how much the project will cost and how long it will take.

Clients expect to receive

- Information from the design team such as concept and design drawings and model information
- Information from the contractor such as the BIM execution plan (BEP), program and cost information, and as-built information and documentation

Clients are expected to input the business case and brief, and any existing information or documents such as survey information. They output detailed requirements in the form of the employer's information requirements (EIR).

Design team needs

The design team is a varied pot of people and can encompass early concept design through to detail technical analysis. Fundamentally, the EIR is about ensuring the proposal and subsequent built asset meets the intent of the client's brief and requirements and that the model represents safe and buildable construction for the next phase.

The design team expects to receive

- Information from the client such as the brief, EIR, and any existing documentation
- Information from the contractor such as the BEP

The design team are expected to

- Output concept and design drawings
- Model information

Construction team needs

The contractor and the supply chain are looking for clarity of technical understanding and scheduling, as well as site analysis in terms of site delivery and quantities and adding as-built information and changes to the model to generate an as-built model for handover.

The construction team expects to receive information from the design team such as technical detailed design drawings and model information. The team is expected to input technical coordination and output as-built information and record information.

Handover needs

What used to be an operation and maintenance manual is now covered by the content of the model. At the handover stage, the client, end user, and facilities manager need information so that they can successfully run and maintain the built asset. The information may be anything about the building's management and maintenance, from operation of windows through to the management systems, heating and energy, fire escapes, egress, and maybe environmental natural sustainability.

Owner-operator needs

Owner-operators are individuals or organizations that own, lease, operate, or control an asset. In short, the same person who runs everyday operations owns the *asset*. Owner-operators are largely using others' outputs, such as as-built information. However, they also need to make changes to the model and update it when something gets repaired or replaced. They need to manage information such as the details of the change and updated warranty details.

Facility and asset management needs

Facility and asset managers may want to compare information across a portfolio of estates. This is especially true of a government client or a university. For example, at a facility and asset management level, personnel may want to consider how a road is performing as part of a network or how a new building is performing as part of the wider university campus.

Re-design/development/refurbishment and demolition needs

The output of the BIM process is an information rich model that can be used in the future. The client may need to refer back to information about a built asset for a variety of reasons, and so having the complete story about a built asset is important.

Projects are a continuous cycle, so they may require refurbishment. For example, a toll road may require a new road surface or updated tollbooths every 25 years, a warehouse building may be converted to residential apartments, or the asset may be demolished to make way for something else. A major problem of sustainable demolition at present is the lack of information about an asset that is to be demolished.

The lack of information about the asset could be because the original drawings no longer exist (or have been damaged over time) or even because the original plans weren't followed in the first place. There may also be a lack of information about the materials used during construction. Having information regarding the materials used during the construction of an asset is particularly important if an asset contains hazardous materials such as asbestos. One benefit that BIM brings is the ability to simulate the construction and deconstruction sequence, improving the demolition process, and all within the confines of a safe, simulated world.

Recognizing who's producing what

Depending on the type of procurement route chosen, a contractor may be responsible for all design work or certain aspects of the design, or have no design work at all. The *responsibility matrix* is a way to define what information is produced and by whom. This document is important because it clearly defines who's producing what information and clarifies design responsibilities. The responsibility matrix should also incorporate information exchange points and the schedule of services. Refer to Chapter 8 for more about the matrix.

The amount and level of information increase over the project lifecycle as the project team develops it. Information needs to be shared during the project timeline among the project team to enable different project disciplines to complete and develop specific packages of work. Information also needs to be shared with the client at specific points along the timeline to enable the client to make strategic decisions. These sections consider the role and importance of information exchanges within the BIM process.

What do you need? Data. When do you want it? Now!

The information exchange element defines when the project team exchanges information. The project team, facilitated by the information manager, exchanges information from the supply chain at defined and agreed points in the process, enabling the whole project team to work from the same base information.

Information exchange points allow for strategic decision points at which the employer makes decisions based upon the facts presented. The sharing of information in this way also facilitates coordination and clash detection to take place. At the end of a particular project stage, the project team exchanges information and data.

We must mention that different people require different levels of information. For example, one client may want everything detailed before going to tender, but another client may only want the minimum information needed so that a contractor can tender for the work and input knowledge and experience.

Exchanging information

Information within an information exchange increases as the design develops. Information exchanges

- Provide the client with an opportunity to review, comment, and approve design as it progresses as well as communicate the design to the project team.
- Allow the project team to submit any statuary approvals or applications and for other third parties and suppliers to pass comment.
- Enable cost estimates to take place and allow contractors and subcontractors to tender for the project.

The exchange of information is an on-going process that's informal. The design team prepares information to allow other members to coordinate information in an ad hoc way. For example, the architect may seek initial comments from the planning authority before any designs are formally submitted.

In the UK, the National Building Specification (NBS) produces a digital toolkit (`https://toolkit.thenbs.com`) designed to enable the project leader to clearly define the team responsibilities and create an information delivery plan for each stage of the project.

Establishing the brief

At the end of the project, the project information model (PIM) is a rich source of information, packed full of data. However, Rome wasn't built in a day, and you don't need all this data from day one. Before you begin any project, you have to be clear about not only who's doing what and how, but also why.

The briefing process and project brief lay down the requirements for the project team. In essence, the *project brief* explains why the project is going ahead and answers these types of questions: Why a new build? Why a refurbishment? Why this site?

Make sure that you have a robust brief in place before starting the concept stage. You wouldn't want to pay for a three-course meal when the client only wanted a dessert! The client undertakes the brief or may draw upon the assistance of professional advisers. Briefly (pardon the pun), you can break the brief down into three stages:

- **Strategic brief:** Addresses the client's business case, core objectives, and reasoning behind the project.

- **Initial project brief:** Further developed within the initial project brief. Consider things such as budget, sustainability, performance and quality aspirations, and project outcomes and deliverables. The client may require feasibility studies to test whether the brief is robust before the design team begins commencing any concept designs. When you have the initial brief in place, you can start the concept designs.

- **Final project brief:** After the design team has prepared the concept design, you must make sure that the brief is updated to include any revisions and updates; after all, design is an iterative process.

Producing feasibility studies

The client uses the *feasibility study* to question whether the project should go ahead, to see whether the project is viable, and to be sure you've looked at all the alternative options. Feasibility studies are prepared alongside the brief and help in the preparation of the strategic brief. Feasibility studies may consider things such as:

- Consents and permissions, such as planning permission and statutory approvals

- Possibilities to refurbish or reuse existing assets and facilities instead of a new-build project

- Procurement routes

- Site assessments and appraisals such as ecology, environmental impact, geotechnical, and contamination

Producing concept designs

The *concept stage* is when the design team members respond to the client's project brief. For example, the client has identified that an asset needs more space to accommodate additional staff following a merger, and so the design team members get to work on planning the initial concept proposals. The concept design determines the basic framework of the project design that the project team will further develop at a later stage, such as massing, structural design, spatial layouts, and environmental and sustainability strategies, and respond to the site's context.

The concept design may include

- Outline project specifications
- Visual indication of proposals identifying key requirements, such as access and maintenance zones
- Information that's suitable for spatial coordination of primary systems or elements
- Schedules of areas and accommodation
- Planning strategy
- Cost plan
- Procurement options
- Program and phasing strategy

Traditionally, the drawing board and freehand sketches have been the tools of choice when getting the creative juices flowing, perhaps because they're quick and assessable. In some situations, BIM tools may be too complex to support free-form modeling because you consider initial shapes and spaces for your concept design quickly, trying new ideas and discarding those that fail at the first hurdle. BIM software is evolving to enable this initial creative phase to occur through the use of concept tools inbuilt to the software that allow you to model shapes and geometry with no restrictions. A number of BIM tools are often marketed as lightweight and intuitive to use, and they allow you to quickly and easily sketch in 3D.

Identifying the Right Information for Initial Users

To get any project underway, certain people within the project environment require certain pieces of information. The decision to build in the first place usually starts with the client instructing the project team; however, the project team needs to understand what it is that the client wishes to build.

The following sections look at the information needs of clients and design teams and how you can help to influence and guide project teams to provide clear requirements and the relevant data as part of coordinated deliverables.

Engaging clients

Many clients may need educating as to what BIM is and the benefits it can bring. Think back to when you made a large purchase, such as a car. Were you interested in the process that the car went through during the design and manufacture process, or were you more interested in the specification, the price, and whether or not it could get you from A to B and fulfil your requirements? Remember, most clients aren't interested in the process, only the end result.

A client may be embarking on his very first project and as such may require the project team to advise on the type of information he should be asking for. Perhaps this project is part of a number of projects the client has. Regardless, the client still requires information at strategic points in order to make informed decisions along the way.

Clients are generally looking for

- Increased building performance, achieved from analysis
- Shorter project schedules
- Reliable cost estimates and cost certainty
- Visualizations, not just for marketing purposes but also to engage other stakeholders and understand the scope of the design team
- Optimized FM strategies — by simulating building usage and understanding energy requirements, for example

Coordinating the design team

Coordination of information is a workflow that requires cooperation from the whole team and preparation and planning from the outset. Unfortunately, as of yet there isn't a magic button that will just coordinate the data for you. Cooperation still requires a good old-fashioned conversation among the project team.

A BIM kick-off meeting is an opportunity before any modeling commences to discuss items such as naming conventions, how to share and transfer information, and any particular software file version requirements. In the section "Recognizing who's producing what" earlier in this chapter, we discuss the responsibility matrix as a way to define what information is produced by whom. (Refer to Chapter 8 for more about the matrix.)

Producing developed designs

The purpose of the design development stage is to produce information that provides a visual representation of the proposals and confirms the brief for the technical design stage, which supports full spatial coordination. Information at a developed design stage should

- ✔ Provide developed principles of the design to a greater level of detail than the concept stage.

- ✔ Provide information that can be coordinated between all professionals.

- ✔ Provide visual development showing coordination for general size and primary relationships between different elements of the construction.

- ✔ Form a brief for a specialist subcontractor or fabricator to progress with technical design, fabrication, and installation. Include critical dimensional coordination, performance requirements, and qualities of finish.

Contractors

Although typically the client engages the contractor after the design team prepares and details the design, getting the contractor in early means that you can benefit from a coordinated and truly collaborative approach, and gain from the contractor's expertise and knowledge. Chapter 16 discusses the minimum information contractors require.

We use the UK a lot as an example, because the UK government has recommended, through initiatives such as soft landings guidance, that the project team considers a building's "in operation" phase throughout the whole project lifecycle. When you establish required performance outcomes and the operational budget at an early stage, you can then compare them to the actual performance outcomes. Performance outcomes can all be enhanced and assisted by early contractor engagement.

Information managers

The information manager is responsible for assimilating, synchronizing, and managing information, and managing the processes and procedures for information exchange on projects. Importantly, the manager makes sure that the fundamentals are in place from the start and that the project team establishes and agrees on an information structure and maintenance standards from the outset.

The common data environment (CDE) is the place in which the project team shares information. A key role of the information manager is the setup and management of the CDE, helping to facilitate data drops. A *data drop* is information that is extracted from the model and then shared among the project team so that the client can make strategic decisions along the way. Chapter 4 discusses the CDE in greater depth.

In the CDE, model information comes together to form a single master, and with the help of the information manager, information is validated for compliance.

The information manager expects to receive model information for the design team. The manager inputs model information received from the design team, which the project team can then validate for compliance with the information requirements. The manager outputs recordkeeping, archiving, and the audit trail for the information model.

The information manager isn't a designer. Although the information manager may have come from or have a design background, this individual doesn't have any design responsibility.

Building Digitally

BIM enables you to add information to the digital model so that you end up with a digital virtual project before you get on-site. Not only does information increase during the project timeline, but also the type of information changes. For example, during the design phase you use models to visualize and show coordination. You then need information for cost and constructability as you move to the construction phase. Then finally, after the asset is constructed you need information about the installed products and systems, and any testing requirements or commissioning certificates.

The following sections look at all of the roles and responsibilities needed to develop, coordinate, and output digital buildings and assets. Some of these roles are purely about modeling tasks, but many are about synchronizing with other information such as specifications and cost data.

As we discuss in the "Changing how the end user accesses information" section, later in this chapter, having data in a digital format brings with it new means and ways to store and archive information so that the client, end user, or facilities manager can easily retrieve it, all of which need to be updated and maintained.

Producing technical designs

The project team develops information further at a technical design stage. Information is developed to provide visual information and the fixed principles of the design that support procurement. Information at a technical design stage should provide

✔ Developed coordination between all professions

✔ Visual representations showing coordination for general size and relationships between different elements of the construction

✔ Dimensionally accurate visual representations that indicate the primary performance characteristics (graphical information represented may alter depending on visual information to be produced, such as scope of work drawings, setting out, floor loading, and so forth)

✔ Typical construction and installation details that are separately produced but linked to model elements and adjacent constructions

Specifying for BIM

A specifier isn't a job title as such, but rather a function that someone performs. You can think of anyone who makes or influences the decision to use a specific building product or service as a *specifier*. That list could include the architect, surveyors, engineers, contractors, interior designers, facilities managers, related professions, and even the client.

Traditionally, construction professionals with significant experience undertake project specifications, so they know what products to select or specify. Specifications often commence after the detail design is complete and perhaps in some cases a little late in the day. Because the specification is often the last task completed before a project goes to tender, it can be rushed as Friday afternoon rolls around. Through BIM, the industry is not only seeing a new generation of specifiers being empowered, but also project specifications taking place at a much earlier stage in the process.

Chapter 9 discusses the digital plans of works (DPoW). Plans of work have been authored to take the BIM process into consideration where more information is produced earlier in the project lifecycle. Having earlier information gives you the opportunity to start to specify at a much earlier stage in the process; for example, you can consider performance requirements.

Synchronizing with other sources of information

Through the use of BIM processes, you can connect BIM objects to the specification. Some specification software also allows synchronization between objects and technical guidance and standards, at the point where the designer most needs them. *Synchronization* is connecting a geometric BIM object to nongraphical information, such as a specification clause. Furthermore, if the designer is a subscriber to services such as the construction information service (CIS), then the designer can download instantly any technical documents cited in the specification that are available. Refer to Chapter 10 for more about connecting BIM objects to other sources of information.

Producing construction developed designs

The project team updates information during the construction process to reflect the final design and to provide a future reference to sit alongside the operational & maintenance (O&M) manuals. Information at a construction stage should provide

- ✔ Developed and coordinated visual information to provide full information to support construction or installation

- ✔ Visual representations showing final coordination for size and relationships between different elements of the construction

- ✔ Graphical representations that are dimensionally accurate indicating primary performance characteristics and sufficient information to support installation

- ✔ Typical installation details separately produced linked to model element and adjacent constructions

Bringing in BIM managers

All the different job titles — BIM manager, VDC manager, BIM coordinator, BIM engineer, BIM modeler, BIM technician, BIM specialist, BIM consultant, BIM information manager — can get a bit confusing. Before we jump the gun and dive into defining the different job titles and descriptions, take a moment to consider the needs of your business and what you're trying to achieve. Depending on the scale of the organization, many of the roles in the following sections may overlap in parts or even focus on a very particular aspect of the role if working on a large-scale project.

In determining competency, many people turn toward qualifications and certification schemes. Increasingly, colleges and universities are including BIM modules as dedicated BIM courses. A number of professional institutions and organizations offer certification schemes leading to entry to a maintained register. These schemes vary in entrance requirements, from providing a CV and BIM case study providing evidence of core BIM competencies through to more formal educational routes where you sit an exam at the end of the course.

BIM managers

The humble manager who looks after the company's IT, software, hardware, and CAD standards is a fairly established role. Many BIM managers are former CAD managers with a new title. However, to be a successful BIM manager, you need a good understanding of BIM principles and how they affect your project.

The BIM manager should be a go-to person for those tough questions as well acting as a trouble-shooter and facilitator. The role of BIM manager, however, can cover a wide range of tasks, such as developing company policies, processes, and protocols, as well as advising on strategic issues such as change management. Like the CAD manager role, the BIM manager needs to be well versed in current and emerging technology trends and software, and is responsible for hardware and software installations and upgrades, or liaises with an IT department if the organization has one.

The BIM manager sets up any project templates and establishes model and company BIM standards and protocols. The responsibility of training in both process and software may be delivered or facilitated by the BIM manager, and he may also look at wider strategic decisions such as human resources and task allocations. The BIM management role may even be merged with a director of BIM role. Think of the BIM manager as a guide who helps the team members make the right decisions.

Typical duties that the role of a BIM manager may undertake include

- Marketing, including promotion of BIM services on the web, in articles, and at conferences
- Managing, including BIM object management, object creation, and approval quality assurance (QA)
- Documenting processes, policies, and procedures and enforcement
- Handling interoperability issues and data exchange
- Training staff and setting up any test requirements for new recruits
- Speaking at or attending conferences and workshops and, more importantly, bringing back any learning to the company
- Being on hand to answer questions from colleagues and provide general support
- Evaluating, recommending, and upgrading software when new releases and updates become available

BIM managers require a good understanding of

- Construction
- IT and software
- Project management

Be careful not to just take an existing legacy role from a CAD environment and put the word *BIM* in front of it. The definition of any role or responsibility should take into account your BIM strategy, goals, and vision.

BIM consultant

The BIM consultant works with organizations to support the successful adoption and implementation of BIM. Organizations may use the services of a consultant to bring in a particular expertise. Because the discipline of BIM is wide ranging, so too is the scope of services a BIM consultant offers. The services may

- ✔ Focus on a particular area like strategy, looking at BIM adoption and future visions and aspirations
- ✔ Be functional, such as producing the relevant documentation, processes and procedures, and policies in accordance with the company's strategy
- ✔ Be operational, whereby the consultant performs the process of implementation

Although BIM consultants may suggest various BIM tools and applications, they don't offer or sell software products. So they must remain independent and impartial with regards to a particular platform. However, operational consultants are likely to be affiliated to a software supplier to help support the successful implementation of products and services.

Reducing issues on-site with clash detection: BIM coordinators

The ability to identify clashes and mitigate them before they're constructed results in fewer clashes on-site. *Clash detection* should be an on-going process, and the information manager needs to coordinate models and documents between the whole design team throughout the design process. This is enabled by a willingness to share and collaborate so that you can reuse information and keep rework of information to a minimum. (Refer to Chapter 16 for more specifics about clash detection.)

The BIM coordinator leads his construction discipline within the project and coordinates any clash detection and any actions or outcomes that may be required. You may also recognize this role as a digital project coordinator, BIM leader, or BIM integrator. Often BIM coordinators are construction professionals who are focused on a production role but have found themselves as BIM coordinators, perhaps because of the knowledge they've acquired, and have naturally drifted into this role, or maybe it has been developed or created around them.

Of course the duties, roles, and responsibilities of a BIM coordinator differ from organization to organization and on project to project, but in the main you can think of BIM coordinators as involved in the production side of BIM. This role has a wider set of responsibilities than information manager, and BIM coordinators usually coordinate their work with the entire design team.

Sometimes people use the terms BIM coordinator and BIM technician interchangeably. The BIM technician may also be known as a BIM modeler or BIM designer, and is more likely to be working in just one BIM platform and on one particular discipline.

Authoring objects and models for BIM

In order to have information to coordinate, you need the individuals who are focused on producing the information.

You can obtain BIM objects from a number of different sources, including BIM object libraries, public BIM library portals, and private BIM library portals, or you may create them in-house (see Chapter 10). In order to produce good-quality BIM content, the BIM author has to have a good overall understanding of BIM software and how it's used within the project environment.

A BIM author may create content for a design team, manufacturer, or BIM library content provider. In the same way that BIM managers and BIM coordinators need to know current industry trends, BIM authors should attend conferences, research, and gather information to fully understand how others, like facility managers, use content. You only have to think about the development of the smartphone and the fact that the iPhone didn't exist before 2007 to appreciate technology is moving at an alarming rate. Therefore, BIM authors have to consider not only current trends and standards but also those that are emerging.

Strengthening the Contractor Supply Chain

Building stronger and more collaborative supply chains increases the value derived from products and also attracts investment. A strong supply chain is good for the economy as it

- ✔ Helps guarantee security of supply
- ✔ Makes advancing technology easier through collaborative innovation
- ✔ Enables faster response to changing markets and customer needs

Other sectors such as oil and gas, automotive, and aerospace have mapped their supply chains, identified strengths and weaknesses, and prioritized areas for development.

The next sections go through the process of increasing the amount of information available, from concept design right through the project lifecycle. You need to be clear how the project team wants to access data and what that means for the way information is transmitted to the supply chain. You need to adapt some of your project information for different audiences, from clients to BIM specialists.

Evolving from the concept design

At the concept design, outline proposals for the architectural, structural, and building services designs are prepared. Outline specifications and preliminary cost information are also considered.

At a discrete point a concept sketch becomes a generic placeholder object. A *placeholder object* is used when the project team has yet to decide upon the final project or solution. The object then evolves into a manufacturer object (also referred to as a *product object* or *proprietary object*) that represents an obtainable product provided by a manufacturer or supplier.

Developed design to as-built

The developed design eventually progresses to represent what has been constructed. This consists of a mixture of as-built information from specialist subcontractors and the final construction issue from the design team. The client may also wish to verify as-built information by using laser surveying technology to bring a further degree of accuracy to this information.

The object finally becomes an as-built asset object with geometry and data reflecting the actual installed product. Not all information lives in an object, as we discuss in Chapter 10. Although you can contain information such as product reference numbers within the object, sometimes linking to information outside of the BIM is better; for example, linking to a user manual through a simple hyperlink.

Table 17-1 considers the typical level of information that is required for each project stage.

Table 17-1	Levels of Information
Description	*Example*
What is typical for concept stage?	A simple description outlining design intent
What is typical as the design develops?	The specified overall performance of the deliverable
What is typical in technical design?	The prescribed generic products that meet the desired overall performance requirements
What is typical in the construction phase?	The prescribed manufacturer products that meet the generic product specification
What is typical for operation and maintenance?	The key properties to be transferred into an asset database

Changing how the end user accesses information

Although BIM is full of useful, relevant information, it's not that useful or even relevant if the end user can't leverage the information from the model at the end of the day. Knowing how the information put into the model will be used in the future and by whom is impossible. That's why the information should be in a structured and open format.

For example, you may know the weight of a building component, so you can include it within the model. This piece of information may be useful in the future for someone who has to return a faulty piece of equipment to the supplier — he can now work out any shipping rates.

Leverage information from a BIM for use in facilities management may require moving to specific BIM facility management tools or to a third-party BIM add-on tool.

Transmitting information

Many people within the process require BIM data for different things, and this information passes from one person to another: from contractors and design teams to the clients and owners, to the day-to-day facility managers, to the window cleaner, to the replacement carpet-tile manufacturer.

Assume a federated model exists somewhere. Now you can generate information such as drawings, specifications, documents, and reports from the model if the information is both structured and in an open format. You access this information directly via the model or you export information. You can communicate and transmit information by

- Exporting to portable document format (PDF)

- Exporting to extensible mark-up language (XML), for use in a database or spreadsheet

- Taking a screenshot and sending via email

- Accessing information on a mobile device through a BIM viewer

- Exporting to a text file or cutting and pasting text

- Exporting to COBie/IFC, which you can then put into proprietary FM software

- Printing out on paper (although the industry is slowly moving to a paperless environment)

Consulting with specialists for BIM

A *specialist* is someone who concentrates or focuses primarily on a particular subject or activity, and that person is likely to be highly skilled and knowledgeable in a specific and restricted field.

A BIM specialist may be an academic or a researcher who devotes his time and interest to a particular aspect of BIM. For example, he may focus on the development of BIM applications and technology, or perhaps the development of open standards. A BIM specialist could equally focus on modeling components, software development, or analyzing data.

Managing Built Assets

The operational and asset management benefits of BIM are significant. In fact, if you compare the cost of a building's construction to its operational life costs, the construction costs pale into insignificance. You see that BIM is partly about embedding, linking, and referencing data in the model at early stages of the project. The inclusion of this information may only pay off many years in the future.

The client is procuring a digital built asset as well as the physical one. This digital asset becomes a tool for the long-term management of the facility and, if it's kept up to date, it forms a live record of the building at any point in time.

The following sections split the overall concept of asset management into a number of key ideas, including the maintenance of digital information in the form of an asset information model (AIM) and how facility managers benefit from access to BIM deliverables.

Maintaining asset information

During the process of design and construction, lots of information, documents, manuals, and certificates are produced, and the project team needs new ways to collect, filter, and update them. Many basements, lofts, and cupboards across the world have in them neatly rolled-up drawings and three-ring binders lining the shelves. Imagine that over the weekend while the office was closed some lowlife tried to break in and damaged the back door in the process. Your role is to find out who manufactured the door and where you can source a new hinge to replace one that was damaged. Perhaps needle and haystack come to mind!

Chapter 8 discusses how the project team transfers information from the project information plan (PIM) to the asset information model (AIM). The project team records the requirements for an asset information model within the organization information requirements (OIR) and asset information requirements (AIR).

Managing assets and BIM for FM

In Chapter 9, we discuss the suite of documents that you can use to help explain what BIM Level 2 means. One of these, PAS 1192-3, looks at operational phases of construction projects, and specifies how the project team transfers information from the PIM to the AIM, and how an AIM is generated for an existing asset. Of equal importance is how the project team then retrieves and passes information on to an existing enterprise system such as a database.

Chapter 11 specifies how to work with COBie information. The purpose of COBie is simple: it's a way of capturing critical information for owners and operators to assist with the management of their assets.

The project team shouldn't duplicate information and data. In some cases linking and cross-referencing the AIM to an enterprises system may be better. Integrating information means the facility manager doesn't have to re-enter information, resulting in better quality of data because it's more accurate.

For example, COBie information increases during the project timeline. If you consider a boiler, for example, the project team captures COBie information such as the manufacturer name and product name. You can then extract this information from the BIM (either through a COBie import or through direct integration with the BIM platform).

The interoperability of data is critical. You can't guarantee that a proprietary format and the technology that supports a particular software application will be supported in the future. Open-source data formats such as COBie and IFC safeguard access to data in the future. Think about an asset manager for a university or a city district. He may have five projects, two of which were created in Autodesk Revit, two in Bentley, and another in Graphisoft ArchiCAD. The costs are in Cato, Exactal, or Calcus and the specification was produced in ARCOM, eSpecs, or NBS Create.

COBie is a way of presenting your data and isn't a request for new or additional information other than what you currently produce for a project.

Making BIM work for FM

As-built or the as-constructed BIM serves to document what was actually constructed, so the client, end user, or facilities manager can reference it in the future for projects, maintenance, or operation and management. Nongraphical data is usually better accessed and updated in a database through the use of unique identifiers.

FM is clearly where clients are going to realize a lot of the performance, maintenance, and energy benefits, but remember that the value of information decreases if it's not kept updated to reflect any changes within the physical built asset.

Part V
Exploring the Future of BIM

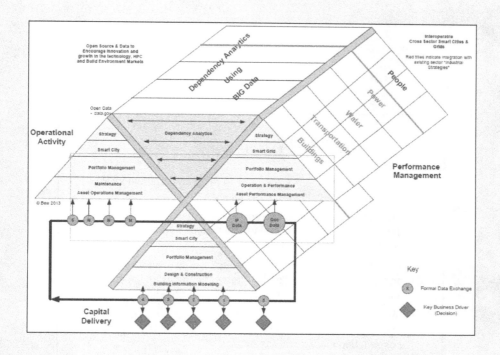

In this part . . .

✔ Look at how new technology is disrupting construction forever and how you can be part of the change.

✔ Find out what is limiting current levels of collaboration and prepare yourself for where BIM will go next and for new frontiers in building information.

✔ Take BIM to the next level by connecting it with new business and urban technology like the Internet of Things and machine learning to empower your business with information.

✔ Overcome the status quo and appreciate that emerging roles and modern working practices are changing the face of the construction industry for the better.

Chapter 18

Building the Future of Construction

A s we mention several times in this book, construction can be pretty slow to change. For many decades, the industry has seemed to resist revolution and found comfort in its long-standing traditions. Now, though, some amazing technologies are finally making an impact, sometimes years after other sectors. This relatively gradual adoption of new ideas is actually not such a bad thing, because it helps those changes to stick around for the long term.

Part V of the book contains three chapters looking at the future of construction, BIM, and other new technology. This chapter specifically addresses the future of construction. Although current standard practice is likely to continue on many projects, someday the way you visualize a construction site and your built environment assets will be very different to the images that probably spring to mind today. You need to think about how the fast-changing world may impact design and construction, and how you allow innovation to make efficiency happen.

In this chapter, we show you how understanding the future of construction helps you to plan ahead for the long-term impacts of BIM. You find out about disruptive technology, new methods of procurement, and how BIM fits neatly with the movement toward modern prefabrication and off-site manufacture (OSM).

Seeing Trends and Looking Ahead

One of the most interesting things when looking at the history of technology is that people are often caught up in the fastest-moving ideas, right at the cutting edge of what is possible, but major industrial impact is more often the result of a gradual change. It can take a long time to realize the true potential of something that initially gets dismissed as a fad or a gimmick. Here we include some key tips that give you the best chance of spotting the trends changing construction for good.

Thinking about your daily BIM implementation may seem complicated, but considering the future is just as important for your long-term objectives. When thinking about what's on the horizon for BIM, here are three patterns to always have in mind:

- ✔ **Open-mindedness is essential.** By being overly optimistic or pessimistic because of the information right in front of you or because of your personal beliefs or experience, you may totally miss potential alternatives. Try to think of every scenario you can imagine when planning ahead.

- ✔ **Current doesn't mean permanent.** Just because today something is either very common or the focus of people's excitement, no guarantee exists that it won't be overtaken by something newer and be forgotten or fail altogether in a few years. You can easily get fixated on what already exists and current limitations. The future is complex.

- ✔ **Trends don't last forever.** In the same way, new technology trends may seem like they'll last for many years, but history shows that even things that seem embedded are subject to change at any moment. As the financial world says, past performance is no guarantee of future success.

Getting caught up in the hype of BIM as just another industry buzzword is a mistake. BIM isn't just a fad; it's the fundamental set of principles and processes that can allow construction to move into the digital age, whether you want to call it BIM or not. BIM as a process is about defining the level of technical accuracy, consistency, and standardization you want to deliver and using technology to achieve that level.

Your guide to futurism

Perhaps you're wondering why *BIM For Dummies* is trying to predict the future. Well, we're not in the business of fortune telling with crystal balls! What we like to do is called *futurism*. It's like the subject *history*, which observes old events to make sense of the past, but instead studies what's going on now to come up with potential alternatives for what could happen next. This isn't just guesswork, it's detective work. This chapter and the other chapters in Part V give you a better understanding of the ideas shaping the design, assembly, and operation of future buildings and cities. You may already be familiar with some ideas, but we hope to provide lots of inspiration too.

Surveying the Construction Landscape: Overcoming the Status Quo

What keeps the construction industry from thinking about the future and moving forward? One strong explanation is the *status quo bias*, which tends to want things to remain as they currently are. (*Status quo* is a Latin phrase meaning "the state in which things were.") We can certainly say that about the traditional construction industry. Some people have compared the industry to an oil supertanker that takes a long time to change course or turn around. That perception is finally changing, though.

After what has been a tough decade for construction in economic terms, a lot of analysts are forecasting steady and significant growth in the sector. All the data analysts are agreed on one point — that much of that potential growth is related to radical change and digital methods, because they will lead to new efficiencies. Think about how technology you see evolving around you, like new devices and intelligent software, can be applied to BIM.

The following sections show you how those traditions that have existed for a long time are finally changing, as construction industry leaders learn from digital innovation happening elsewhere. These changes will have an impact on the *demographics* of the industry, which just means the characteristics and backgrounds of the people around you. You also see how BIM has the opportunity to have as much impact on construction as sustainability awareness has over the past few decades.

Changing the face of construction

The built environment industry carries a few old stereotypes about the type of people you encounter, the type and quality of work that individuals do, and their experiences. Many of these stereotypes have been very accurate in the past, but the advent of BIM will result in totally new roles and modern working practices. It will also encourage new, younger graduates from subjects like data science and robotics to approach construction with totally fresh eyes. This section looks at the impact of those changing groups on the industry.

What do you think of when you hear the word construction? Despite what you already know about the digital future of the industry just from picking up this book, that's not what most people think of. Here are a few common (mis) perceptions:

- ✔ Construction is dusty and dirty work.
- ✔ Construction is hazardous and dangerous.

- ✔ Construction is full of inequality and discrimination.

- ✔ Construction is set in its ways.

An influx of dynamic, inspired, young, and diverse people in the workforce will only occur when industry organizations work together to address some of these key aspects on the ground. Think about how you can effect change in the organization and construction sector you're involved in. Does your work contribute to a positive image of digital innovation or just reinforce the dusty stereotypes?

Look at how the face of construction is changing. You can build more sustainably, safely, quickly, and connectively than ever before. Through innovations like mobile devices linking with near-field communication (NFC) and as pre-fabrication evolves into designing for manufacture and assembly (DfMA), you can reduce or remove much of the inefficiency of traditional practice and rethink the construction site. Working together, everyone can make the industry accessible and welcoming.

A recent report by the Center for Construction Research and Training (CPWR) showed that the median average age of construction workers in the United States has been increasing for years; it was 38 in 2000 and the latest figures indicate that this number is now up to 42.4. The construction industry isn't attracting enough young people and new graduates. Lots of other exciting industries exist, and construction just can't compete with leisure, entertainment, and Internet businesses.

Revolutionizing construction and site processes is going to rely on *soft skills* (the nontechnical abilities people have, like cooperation, problem-solving, and enthusiasm) and how people work and communicate as much as it will rely on new technology. You can find out more ideas for this in Chapter 12.

Automating construction

The construction industry has embraced technology in the past, and just like other sectors, construction is becoming more automated too. Just like a home dishwasher or vacuum cleaner, machines increase efficiency.

You can compare the shift in farming to how construction has changed. In the early 1800s, more than 80 percent of US workers were in farming. The number of agricultural workers has seen year-on-year decline and is now closer to 2 percent. Much of that decline is because machinery replaced manual labor, as it has in manufacturing during the last century.

As we discuss further in Chapter 20, construction is one of the last industries to still be full of repetitive manual tasks carried out by humans. So suggesting that the increased digitization and automation of construction workplaces won't have a negative impact on manual labor is naive, because it probably will. Innovation will lead to new opportunities for the construction industry, which will require everyone to develop new skills and adapt to change.

Watching for encouraging signs: Performing better

Construction can sometimes seem a bit backward. It not only appears to be slow moving; it resists transformation. In reality, construction does change, but that change just happens very gradually. This section highlights how much improvement has taken place recently, in areas like sustainability and site safety.

The three of us are all very optimistic about this industry. Despite how slow people can be to adopt change at times, you don't have to look too far to find examples of how the construction industry demonstrates the desire to move forward. Think about the following:

✔ How hazardous methods have been removed through health and safety regulation and the number of deaths and injuries caused by construction work has reduced.

✔ How people encouraging sustainability has made ecologically responsible products far more common, reduced site waste, and lowered CO_2 emissions. This has also increased the average efficiency of housing and public buildings like hospitals and schools.

In the next sections, you can find out how encouraging people to consider the environmental impact of construction work is a great example of the kind of progress that we want to see happen with BIM.

Sustainability just means looking after a system like the earth's environment, so that it continues to produce enough resources to supply people's needs without running out now or in the future. Over time, construction has gone from being quite wasteful to leading the way. Especially in the last 20 years, the impact of industry on the environment and climate change has been brought to the front of people's minds. Construction and its suppliers and manufacturers have changed the climate. Through sustainable practice, you can begin to slow the damage being done and maybe turn around some aspects altogether.

Sustainability and BIM have a lot in common; just think about the following:

✔ **Knowing where things come from:** Being able to confirm that the forest from which timber was sourced uses ethical principles or that power is drawn from renewable energy schemes is essential to modern construction projects. BIM processes handle information to ensure that it's accurate and responsive to change. A key example is in product information and being able to quickly interrogate the information model to find out all the relevant parameters of an object, including its source and its environmental properties.

✔ **Understanding industry schemes:** To support sustainable sourcing and design, schemes like BREEAM (www.breeam.org), LEED (www.usgbc.org/leed), Programme for the Endorsement of Forest Certification (PEFC) (www.pefc.org), and the Forestry Stewardship Council (FSC) (https://us.fsc.org) are required. Organizations like WRAP (www.wrap.org.uk) offer advice and guidance on limiting waste. In the same way, BIM industry schemes, mandates, and standards are needed to ensure everyone is heading in the same direction. At a detailed level, you need to understand the documents, frameworks, and standards that indicate how project teams can achieve the targets; sometimes they can take a long time to be produced or to be transparent and explained. Hopefully, this book can help you start.

✔ **Changing the culture:** A lot of waste was due to cultural factors; for example, the effort of sorting demolition waste that others could reuse versus just throwing it all into a skip. To change that attitude, site workers needed information on the importance of recycling where possible and the impact on the environment. BIM is going to require organizations to provide the same level of information across all levels of the industry, to educate on the benefits of new BIM processes; otherwise the adoption could fail at ground level.

✔ **Cooperating in design:** The industry knew for a long time that waste was a huge problem. Take, for example, the vast amount of offcuts in plasterboard and timber. Often, this was due to design dimensions that didn't make use of standard sizes and required a lot of cutting on-site. Design and build contracts where contractor teams were brought onto the project at an early stage changed this, because the teams had a much clearer idea of buildability issues. Asking the design teams to keep to standard product dimensions as much as possible reduces wastage. BIM takes this idea a step further, ensuring that the entire team knows the exact quantities and dimensions in the model environment. Buildability issues are displayed clearly, allowing for refinement of the design to reduce waste and make for an easier build.

✔ **Impacting the entire supply chain:** In any construction project, you're one link in a huge chain of processes and people. The nationwide drivers of environmental awareness and personal responsibility encouraged the drive for sustainability in construction. Whether it's wasteful packaging, damaged components from poor site management, or demolition waste, everyone on the project timeline has an opportunity to make a difference. Appreciating that you're part of something much, much bigger is so important in BIM. Everyone is in this together. As procurement gets faster, just-in-time delivery of exactly the required amount of each material will generate huge efficiencies.

BIM is a solution to lots of material inefficiencies and communication problems that result in abortive, wasteful work. In fact, lots of BIM best practice aims to help sustainable construction, including solving the performance gap where real-life performance just doesn't match the estimated performance of the design. Check out Chapter 15 for more on sustainability.

In the UK, the Construction Industrial Strategy includes a commitment to lower CO_2 emissions by 50 percent by 2025. The UK government sees BIM as a key part to making that reduction a reality. Making the construction industry more sustainable didn't happen overnight, and the industry still has a long way to go. Construction remains one of the largest contributors to waste generation, responsible for up to half of what goes to landfill. Change relies on policy, performance, people, and protocols all gradually evolving.

A 2010 report by the contractor/developer Willmott Dixon includes an estimate of global pollution that can be attributed to buildings. The report suggests that 23 percent of air pollution in cities is from buildings and that 50 percent of climate change gases and ozone depletion is due to buildings. That means half of the impact humans have had on the environment is directly linked to buildings. The report also lists estimated resources used in buildings — for example, 90 percent of hardwoods are cut down for use in buildings, along with half of all the water and energy the world uses.

Disrupting an Old-Fashioned Industry

Keeping an eye on new technology, products, and digital inventions so that you're prepared for how they could displace existing processes in the industry is important. Many disruptive innovations can take time to become well known and even longer to become commonplace. If you can look for tested, verified data like examples and case studies, you can adopt the appropriate systems and tools when they're most cost effective and still stay ahead of the curve.

The following sections show you how the future of construction will disrupt tradition and how adoption of technology takes time. It always starts with early adopters — the innovators who lead the way — until there's a tipping point when the majority of people consider adoption as necessary.

Impacting with disruptive technology

The key with *disruptive innovation* is that the existing surrounding industry is affected and eventually replaced by the new idea. The opposite is *sustaining improvements*, which are more like an evolution of an existing technology. Through sustaining improvements businesses compete with one another in making things better and better.

Predicting how long an innovation will take to be adopted can be difficult. Disruptive inventions that change the way people live and work, like the cell-phone or computer mouse, can take around 20 to 30 years to reach common usage. But devices for entertainment or media consumption, like CD players, cassette players, and televisions, were adopted much faster, sometimes in less than five years.

Have you heard the term *early adopter* before? It comes from a man called Everett Rogers. One of his many theories was called the *diffusion of innovations*. He suggested that you could divide groups of consumers into categories based on their willingness to adopt an innovation, and that these groups follow a bell-shaped curve distribution.

This same idea can be applied to BIM by using some common alternative terms for the same groups:

- **Technologists:** When was the first time you heard about BIM or virtual construction? A pioneering few innovators have been trying to encourage new digital systems and processes in a paper-obsessed industry for decades. Often they have just been expensive oddities, like early automobile inventors' attempts to make a steam-powered car.

- **Visionaries:** Some companies have been instrumental in leading the way with BIM, implementing their own way of working, sometimes on huge projects. The rest of the industry was only just hearing about its potential from example case studies and media stories. Computing power was still an expensive investment. You can think of this like the first gasoline cars, which took decades of development before there was really a market for them.

- **The tipping point:** We think that this is where the industry is right now, with more companies seeing a return on investment, their clients imposing requirements, and more countries imposing BIM mandates. This is BIM's Henry Ford moment. Ford was able to use assembly line techniques to make cars affordable. BIM is no longer an eccentric, expensive idea; it's now in mass production.

✔ **Pragmatists:** Reading this book is a great start. As part of the early majority, you're set to take advantage of all the lessons from the uncertain days of BIM implementation and act on your organization's unique situation. This book forms part of the infrastructure. Think of past experience like the improved roads and supply of fuel in gas stations that cars need, along with rules and regulations, which describe the best, safest ways to use a new technology.

✔ **Conservatives:** BIM will become completely commonplace, an assumed foundation for all design activities, site, and construction methods and asset operation. The vast majority will follow best-practice data management simply because doing the opposite means wasting time and money. You'll have as many tools for information control as you have currently for drawing. Just as car producers keep improving performance and providing variety, so the BIM market will respond to inefficiency with new solutions and bespoke options.

✔ **Sceptics:** As with any technology, some will long for the old days and will begrudgingly use newer processes because they have to. Some of the later adopters, especially the very young, may think of a better way of doing things and become. . . .

✔ **The new innovators:** What happens after BIM? What will disrupt it? Nobody really knows! But you can be certain that people will continue to innovate and that technology will always move forward. Just look at the evolution of electric cars and self-driving technology. We predict that BIM will evolve too, and you can read more about that in Chapter 19.

We also cover the BIM adoption curve in far more detail in Chapter 12.

Being mobile, seeing more

One of the major ways that the construction industry has been disrupted has been by increased mobility. Perhaps you carry a smartphone or a tablet and you send emails or make business calls with that device. Maybe you're drawing sketches or writing reports during your daily commute. How do you communicate with your colleagues? Always face to face, by email, or by phone? Do you use video conferencing tools like Skype, GoToMeeting, or FaceTime?

You can quickly see that you don't need to be tied to your desk in order to do these activities. Not only are more people working from home, but you can also send a message from the construction site, from a hotel, or in transit on the other side of the world. Can you see how this frees up the industry to work differently? That's what we mean by *disruptive technology,* things that appear to come out of nowhere and revolutionize the way you live or work.

Just think about what this means for site work. If you see a mistake or an issue crops up on-site, you can quickly call up any aspect of the virtual information model to compare it, communicate it, or change it as necessary. The increased mobility of the industry makes for rapid connections and on-the-fly instructions.

This is a double-edged sword. If someone on the project team needs an answer, you may have to think and respond more quickly than you want to. Don't rush into decisions because you feel pressured. Tell people if you need more time to consider your answer.

In this hyper-connected world, don't forget to have a life outside of work too. At some time, we've all been told off for checking our work cellphones when we should have been on vacation.

Disrupting through investment

Disruptive technology often shuts down an existing part of an industry, but the new innovation quickly becomes a really competitive market full of commercial opportunity. You'll probably have noticed that the BIM marketplace is already full of companies, consultants, and providers who are fighting for the largest opportunities and who see BIM as a potentially lucrative investment.

As part of all this disruption and evolution in the industry, you'll likely see more mergers and acquisitions in the largest contracting, property management and facility management (FM) operation companies. This is a quick way for a company to offer BIM services. More organizations are choosing to grow their offering of skills and the areas they serve by acquiring the existing experts and market leaders.

Just think of the scale of research and development that goes into pharmaceuticals and medicine. A nearly unimaginable level of financial and time investment goes into the development of a new drug or treatment, which is gradually recouped in the cost to the market, whether governments or hospitals. It's not so much of a stretch to imagine one of these new giant super-contractors investing similar levels of money in developing a groundbreaking building method or, through project experience and virtual modeling, coming to an unprecedented understanding about the performance of a particular building type and dominating the market in its construction.

Taking a long-term view

The sheer amount of data being generated by a digital construction industry will reveal a huge amount of lessons in a very short space of time. People will be able to analyze the problems faced by the industry, some of which were invisible to everyone involved. That means some technology will take a while to evolve as it becomes clear where it could help the most.

As we suggest throughout this book, take the long view. In every decision you're making, consider what you're hoping to achieve in the next 5, 10, 20 years as a representative of the construction industry (for example, maybe you want sites to have zero fatalities and injuries), as individual businesses (perhaps you want to reduce your environmental impact), and as an individual (you want to construct a lunar base!).

Approaching a New Era of Procurement

In a presentation on the off-site construction campaign Buildoffsite, Richard Ogden points out that between 1996 and 2006, inflation in the UK rose by 29 percent. The average price of a car in that time increased by 1.5 percent. The cost of construction in the same period? It rose by 89 percent. That figure isn't missing a decimal point: it's 89 percent.

The statistic prompts two questions: why is construction so expensive, and why are cars relatively cheap? Partly, this is something to do with inefficiency and methods that increase the cost of traditional construction, in comparison to advancing safety, technology, and performance in the car industry, the cost of which hasn't been passed on to the end user, or hasn't made new cars unaffordable. We think the industry can work in ways that mirror aeronautical or automotive production, but what does that mean in reality?

In simple terms, it means automating any process that doesn't have to be manual. Sometimes, manual work can result in a beautiful finish, but it's time consuming and resource hungry. Equally, though, manual work can sometimes result in poor quality, so why would the industry not try to program something to repeat time and time again with identical results? The construction industry is experiencing a skills shortage that slows down the entire process.

The following sections explain that getting construction to act more like modern manufacturing is going to require increased use of modular construction, which just means pre-built pieces that you can fit together on-site. Then, you can imagine entire buildings being pre-constructed in an environment that looks more like a factory than a traditional building site and being delivered for assembly like flat-pack furniture. As you make the building process cleaner, the efficiency gains increase.

Providing a modular kit of parts

If you have an interest in architectural history, you may have heard of Georgian-era pattern books. Similar to a catalogue of design options, pattern books were a way for local builders to gain access to architectural design, because the drawings were published for mass readership. These pattern books contained potential designs for various building types, but especially houses, from quaint cottages to grand townhouses and mansions. By copying these standard plans and allowing for a little adaptation, you could almost guarantee quality in design.

Taking this idea a few steps further, what if you weren't just buying plans that were standardized for quality, but could source pre-built modules for entire buildings? The technology for off-site manufacture (OSM) has been around for hundreds of years. Stories exist of houses being shipped from the UK to Australia as a kit of parts and of a whole hospital built in shipbuilding docks and sent to Turkey during the Crimean War. House kits were very popular in the United States during the early 1900s, such as the Sears, Roebuck & Co. model that had 30,000 parts!

You may be more familiar with the term *prefabrication* or *prefab*. Prefab sometimes gets a bad reputation. Generally, people have used it in response to a need for fast housing for huge populations, like during or after wartime when housing shortages occur. The priority was often speed, which meant many were cheap and poorly constructed, resulting in leaks and thermal inefficiency. In the UK in 1968, a tower block partially collapsed after poor construction gave way in a gas explosion. Sadly, those memories linger for many people and a cynicism exists about prefabrication.

Now, though, the very traditional industry we keep talking about is changing and beginning to embrace the idea of modern construction technology, helped by the new virtual modeling techniques to perfect OSM in a factory environment.

Designing for manufacture and assembly

The factory environment of other clean manufacturing industries is exactly the kind of image that promoters of OSM believe can revolutionize construction. From precast concrete units of any shape to entire timber-frame wall elements with pre-fitted windows, doors, and interior finishes, the 3D information model contains the need for data fabricators and contractors to produce accurate replicas of the designer's intentions, sometimes directly milling, cutting, or manufacturing the components from the design model. Where more accurate fabrication data is required, systems can coordinate with robotic machinery to communicate precise detail. The best success seems to come from using the accuracy of the manufacturing process as a driver for better design.

Boothroyd Dewhurst, Inc. trademarked the term *designing for manufacture and assembly (DfMA)* to describe this refinement of design prior to production, but it has become commonplace terminology in modern construction. By optimizing the design of an output, you find its manufacture, assembly on-site, and eventual maintenance or deconstruction much simpler than with traditional methods. DfMA aims to do the following:

- ✔ **Reduce manual labor.** The industry still makes a lot of people do physical, manual work that leads to injury or long-term health problems, including disability. Robotics and assembly line machinery can do hazardous or difficult jobs instead of requiring people to do them. Some projects have reduced on-site labor by more than 80 percent.

- ✔ **Reduce emissions and increase sustainability.** By designing to reduce waste, the assembly line uses only the necessary amount of any material. Project teams can optimize sizes to any dimension and assemble components to fit perfectly. BIM considers the environmental impact throughout the lifetime of a component. Designers can work with fabricators to test performance of building products exactly, unlike in site work, where the reality of untidy work may not be capable of matching the expected thermal or acoustic design properties. Imagine building a door in a factory, where any gaps and airflows can be precisely sealed up by machinery, in comparison to making the door on-site with manual tools.

- ✔ **Increase quality and consistency.** Site processes are typically difficult to replicate time after time. Assembly line and factory technology reproduces exact copies of a design millions of times over. Think about the unprecedented levels of specification that DfMA can guarantee. What does this mean for quality assurance? Over time, you'll notice an effect on warranties and insurance. Also, if you encounter problems on-site, you improve the design and every future component benefits from the change.

- ✔ **Increase speed of production.** With assembly lines you're not reliant on one aspect of a job to be finished before starting your own. Each one of the many components in the factory can be at any stage of the assembly process, so a constant flow of completed products exists. Not to mention, you don't have to worry about the weather! On-site, the speed of construction also increases. Some reports suggest savings of up to 50 percent on project schedules by using OSM systems.

- ✔ **Reduce risk of injury or accident.** An increase in reliance on machines to do unsafe work reduces the danger for site teams. See the later section "Making construction safer" for more information.

One of the key aims is to reduce the number of parts necessary to achieve something. The best example is instead of having to install six batteries to power a radio, have only one battery pack containing the equivalent power. For buildings, this is like having a bathroom pod that contains pre-installed units and basins to be dropped into the building whole, requiring only plumbing connections on-site. An astonishing example of this is the pre-fabricated buildings for new McDonald's restaurants. Using pre-prepared ground foundations, an entire modular McDonald's was recently installed in just 13 hours from start to finish.

You can recognize the benefits of OSM only if the installation on-site is successful. In general, you need specialist teams to guarantee precise execution of the works. Alternatively, traditional site workers need to be trained in new skills to deal with pods and pre-fabricated elements.

The UK is considered to be leading the OSM industry, and one of the companies immediately seeing the benefits of DfMA is Laing O'Rourke, through their *Explo're* OSM facility.

OSM and DfMA won't necessarily be able to replicate all traditional site processes and construction methods. Many of the early examples are concrete based and form standardized components like columns, beams, and planks. Replicating and automating brick or exposed-timber construction to a high quality can be very difficult.

Managing new site processes

DfMA and OSM aim to reduce the scale and schedule of on-site construction. If you can achieve more in the clean, quality-controlled factory, then you spend less time on dirty sites at the mercy of the climate.

Another significant benefit of OSM is the reduction in needing to transport raw materials. OSM means that you don't encounter the awkward logistics of moving steel beams and long pieces of timber around a construction site; you can use manageable pre-built components.

Chapter 3 discusses how radio-frequency identification (RFID) tags can be used to track the delivery, location, and use of every single component as an information model. As this technology becomes more commonplace, contractors will be able to understand the supply and demand of construction materials like never before. Real-time data will increase, informing decisions and accelerating reactions.

By using the rule-sets developed from designs, the transactions between supplier, contractor, and client will become instantaneous. The recent BIM 2050/CIC report "Built Environment 2050" calls this "nanosecond procurement." This has a huge impact on the industry's ability to deliver buildings of exceptional quality to a precise schedule.

Making construction safer

Many site activities are very dangerous, so if OSM can bring those hazardous methods into the controlled factory environment, construction becomes much safer. Having to conduct dangerous activities at height and in high winds, wet conditions, or extreme temperatures makes them significantly worse.

Improving the quality of construction also reduces risk to the end user. Because a contractor can sign off the end product before leaving the factory, testing and commissioning to ensure user safety in the completed facility becomes much easier.

Don't forget that construction becomes safer as soon as the industry starts building digitally. Understanding or interpreting potential hazards and risks from paper drawings and printed Gantt charts can be difficult (refer to Chapter 16 for more information). Having the information model means that the project team can interrogate it in 4D (scheduling) to watch for potential clashes in activity, such as working with heavy elements at height while work is taking place on the ground below, or seeing where harness and fall protection will be necessary. Some also say that BIM's greatest legacy will be reducing site deaths to zero. See Chapter 16 for much more on BIM for health and safety (Stefan's favorite subject!).

Chapter 19

Eyeing the Possibilities: The Future of BIM

In This Chapter

▶ Considering the future of collaboration and data

▶ Building simulations and augmenting reality

▶ Maintaining the fundamental principles of BIM

▶ Keeping your information secure in a digital age

*W*hen you look at BIM, what do you see? You're in the middle of a *paradigm shift* (something that profoundly changes the way people understand things) in procuring buildings; everyone is getting their ideas off the paper page, and the majority of companies involved in buildings and infrastructure have spent the last few years embracing digital technologies. Digitization is having a real impact on the financial and environmental cost of the built environment. The previous chapter looks at where construction is heading; this chapter focuses more on BIM and virtual building information.

If you think about the BIM maturity model (much more in Chapter 7), then you can look at what happens after the industry reaches Level 3 maturity. Level 4 is about behavioral change, and you can argue that the point of doing BIM is really to change people's lives. A very grand statement, we know, but we hope you'll agree with us when you see what we mean.

The main issue with this kind of speculation about Levels 3, 4, and beyond is that they're going to rely on some technology that isn't available yet: either ideas in beta development or one of many potential solutions for how people will work together. Thanks in part to what BIM has already achieved, the industry is leaner, more streamlined, more efficient, and more innovative than ever before. This chapter considers that BIM will continue to evolve and really begin to change things.

Thinking about Collaboration

When history looks back at this era, it may well be called the Information Age. Recent innovations in science and industry have placed high importance on cross-examining data and the same applies to construction. As we discuss throughout this book, software vendors sometimes use BIM as an analogy for 3D modeling, when it isn't that at all. True BIM needs information at its center, and you need to be able to understand the data and call it up on demand.

Collaboration is a funny word. It means working together. We know one prominent BIM advocate who likes to remind everyone that it can also mean to secretly plot an unlawful conspiracy! A better word is probably *cooperation,* working toward a common goal. But among the technologies that promote cloud environments and most of the people who talk about BIM, collaboration is king.

We talk about biases in Chapter 18, and another word to describe how human thought processes are quite predictable. It's *heuristics,* which basically means that people prefer trial and error, guesswork, and so-called common sense to reach conclusions, instead of finding an accurate solution, which can take a long time.

BIM requires a higher level of accuracy and for the users of BIM to clearly understand the exact impact of their decisions on the rest of the project team. It takes time to move from heuristic ways of thinking to an accurate, considered approach that looks at how everything links together and not just your part of the puzzle.

The reality is that you have to think in a completely different way if everyone is going to collaborate well together. As the phrase goes, "You can't collaborate on your own." The following sections look at what stops people collaborating and how BIM will lead to behavioral change in your team, the industry, and, eventually, the users of your built projects.

Recognizing what's limiting current levels of collaboration

Here are four things that can often affect your ability to collaborate with project teams:

✔ **Closed data:** Have you ever seen an "Access Denied" message pop-up? Restricting access to data that should be shared makes BIM processes difficult to achieve. One of the biggest barriers to collaboration is that one generator of information, like a landscape designer, doesn't have access to the data of another, like an electrical engineer, on a project that requires outdoor lighting design. Perhaps they're using obscure file formats or only providing locked-down, PDF-style outputs. Agreeing levels of permission for project teams and creating your information with other users in mind can be powerful.

- **Security concerns:** Maybe you're involved in top-secret projects full of spies in dark glasses and long coats, but even just on a regular project you probably have access to pieces of sensitive company data or people's personal information. Thus, you want to be confident about the security of file storage or file transfer systems any one person on the project is using. Work with your IT teams and digital security experts to develop a strategy that protects commercial interests but also allows appropriate levels of sharing and access. The UK document PAS 1192:5 is a useful foundation for this process.

- **Attitudes and personalities:** People can be awkward about sharing data. Maybe they think a commercial advantage exists in protecting their content, but actually the awkwardness can come down to fear of weaknesses or poor work being found out when they display it to the project team. Also, as Chapter 12 discusses, it would be great if everyone involved in projects was keen, dynamic, and enthusiastic, but you find (at least) one or two grumpy people in the construction industry. Make your project team meetings a place where people feel comfortable that they can bring their problems and questions as well as their expertise and solutions.

- **Collaboration tools and technology:** Sometimes the tools for collaboration simply aren't available, or worse, aren't used. Currently, the most common tools don't integrate very well. However, we think this is probably the least likely reason that collaboration doesn't happen. Blaming technology is easy when people are actually the problem. Selecting tools that have the highest chance of working together is a good idea. Related to this is the need for forms of procurement that encourage project leaders to engage a full team including contractors and specialist consultants as early as possible.

Looking at Level 4 BIM

The industry is close to agreement on what forms the Level 2 suite of standards for BIM compliance. Chapter 7 looks at the BIM maturity model and at Level 3 in particular. But what happens after that? Understanding that BIM maturity also has the potential for Levels 4 and 5 is fundamental to realizing the future of BIM. We provide more detail in this chapter, but in simple terms the BIM levels can be described as follows:

- **Level 2:** Coordinated digital information, but generated in separate software.

- **Level 3:** Increased integrated sharing of data using collaborative software, leading to added value and cost/energy savings.

✔ **Level 4:** Behavioral change: using BIM to influence the real-world impact of your projects.

✔ **Level 5:** Projects that form part of smart cities/grids, providing real-time information to operators and users.

Building Digitally

In Chapter 4, we introduce the phrase, "Build it twice: make the mistakes digitally and then in reality." What this means is that you're building digitally first to reduce the amount of problems you encounter on-site and in operation. Now think about how that process of building digitally can evolve in the future. Figure 19-1 shows how changing technology will impact digital construction. Consider that the levels of BIM maturity are about increased digital sophistication:

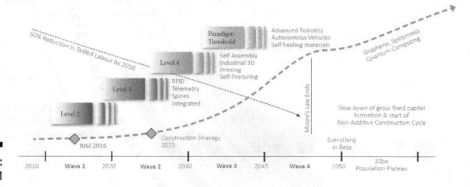

Figure 19-1:
Future BIM
maturity
from BIM
2050.

Illustration by David Philp/Neil Thompson

✔ **Level 2:** Build it digitally on your own, with standard exchange formats to share it and federation to bring it together. The industry is working here now.

✔ **Level 3:** Build it digitally in an online common data environment (CDE), sharing and collaborating with other people. This is the aim for the next five to ten years, with radio frequency identification (RFID) tags and building sensors feeding into this wave of new construction technology.

✔ **Level 4:** Build it digitally including operational data with behavioral feedback. The industry is learning from other sectors about industrial-scale processing, manufacture, and delivery. The future of construction involves self-assembly, robotics, and increased reliance on computer systems and automation of building controls.

✔ **Level 5:** Build it digitally, connected in the Internet of Things. The aim of digitizing the construction industry is eventually to deliver the concept of smart cities where the built environment is fully connected and its data can be interrogated and analyzed by owners and users.

In this book, we don't suggest that one particular software or technology solution is essential to your work. We simply point them out when we think they're very useful or particularly cool, but they're never fundamental to BIM. BIM is always a process, not a technology, and many ways to collaborate digitally will always exist.

This is why interoperability is so incredibly important. (*Interoperability* just means your software outputs can be shared or used by other software, generally because they use open formats. Chapter 10 provides more information about it.) Soon, you'll be using online CDE with federated information, and eventually you may be using collaborative, cloud-based workspaces. But in the meantime, you need to use BIM processes to ensure that everyone on the project team has access to the data they need, even if people are using different file formats.

There's no doubt that this process of digital building is going to become even more essential to the industry. It will continue to develop so that you can simulate exactly how your project is going to be used before it's ever built, and use virtual reality (VR) and augmented reality (AR) to inform and explain your project to clients, site teams, and users. The following sections look at this exciting evolution in more detail.

Digitizing the world

When you're in a meeting, what do you make notes on? Are you using a pen and paper? A laptop? A tablet? Some industries have been transformed by *digitization* (the move from analog forms of data and media like printed books and handouts and handwriting to fully digital, electronic processes that can be connected together into a vast network of online transactions), such as the finance sector, where analog processes like ledgers and checkbooks are nearly extinct. In other sectors, like media, although digital is clearly the future, paper holds on for dear life, and people still buy newspapers and magazines, full of brochures, coupons, and leaflets. Enough of a market exists in the non-digital realm to keep producing paper products. The physical books market remains strong alongside a growing e-book and digital publishing industry. All industries are digitizing, just at different rates. The development of digital economies varies across the globe, too.

A report by Strategy& (part of the professional services group PwC) ranked different industries by their levels of digitization. In sectors where it was relatively simple to move data inputs and outputs into the digital world, like finance and media, digitization has been underway for decades and generates new efficiencies every day. However, the construction industry and the hospitality sector are at the bottom of the pile. These industries are seen to be very labor intensive with many analog processes still in existence.

In order for a digital economy to work, you need certain things in place:

- ✔ A physical infrastructure for digital: the cables and pipes, power, and networks.

- ✔ A regulatory infrastructure for digital: the standards and agreements, security, and safety.

- ✔ A workforce ready for digital technologies: Generation Z + Generation Alpha. These terms are popularly used to describe those born (approximately) between 1995–2009 and after 2010. Generation Z is the first generation to have never experienced life without the Internet or digital media and so is more equipped to access the information it needs than any previous generation in history. Generation Alpha is the next generation and the first to have never experienced life without social media, mobile devices, and collaborative education. This group is sometimes called *digital natives*.

Simulating and augmenting reality

Interestingly, you can discover a lot about real-world human behavior, psychology, and decision-making by looking at online communities. In games like World of Warcraft and League of Legends, developers bring in economists and psychologists to make the experience more realistic. Academics have used the glass-box, laboratory-like world of online games as a rich mine for social study. Recently, US media scrutinized Facebook for a series of psychology experiments it tried with hundreds of thousands of users, artificially changing news feeds to show happier or more miserable updates and seeing what effect it had on users' own posts.

Simulation — and the power to run multiple versions of similar computations over and over and see the results – is one of the main advances of BIM. Interestingly, many of the online games use many separate servers, which means the makers can measure each of the behaviors in unique situations and apply various changes to affect only one set of users.

The following sections examine how you can apply simulation to the built environment and how modeling simulations, including augmented reality, can optimize the performance and construction of built assets.

Examining simulation

Working in a trial-and-error way simply isn't possible for many industries, such as aeronautics and space exploration, invasive medical procedures, or emergency military equipment. In these scenarios, digital simulation not only refines design but also saves billions of dollars. The built environment is complex; it's a huge consumer of energy and is embarrassingly accountable for global emissions.

Simulation can help to improve performance and resolve those difficult operational problems that end up being really costly to change post-construction. Not only that, but you can explore thousands, even millions, of potential alternatives in a very short space of time. You can totally optimize the design so that the finished end product has the best possible chance of achieving or exceeding expectations.

Here are some ways to use simulation:

- ✔ **Building performance:** Energy modeling or comparing alternative lighting, heating, or cooling systems.

- ✔ **Operation and emergency:** Simulating typical human patterns and interactions to see how people move from rail or subway platforms, or exit a building when escaping a fire.

- ✔ **Climate models:** Assess how climate change, super storms, or rising water levels will affect the project's future.

- ✔ **Durability in use:** Simulating things like what the building will be like in five years, how quickly those trees will grow and shade the area around it, and how much traffic can pass on the bridge before the road surface will need replacing.

Energy simulation modeling is an obvious added value of BIM implementation. By improving your digital representation of *accurate geometry* (connections, joints, and material behavior) and understanding the data behind the objects, you can simulate energy performance more accurately than ever.

That's not to say that it's an easy task. Traditionally, energy-modeling tools have been very specialized or complicated — running secret calculations for heat loss versus overheating and taking weeks to generate rainbow-colored lighting models. Well, that's not the case anymore.

The two main innovations are speed of feedback and real-time data. Instead of needing to run separate software or take the model off-line, you can simulate it natively in many BIM platforms. You can also optimize your design on the fly because the model immediately updates anticipated energy calculations based on your changes. You can use real-time climate data instead of out-of-date averages too. (Examples are Autodesk Formit and Sefaira.)

Establishing performance-based requirements

A popular idea about BIM's future is that clients and owners will no longer just be looking for vague objectives or aims about a project's performance. Instead, clients will ask you for project outcomes to form part of the contract.

Imagine a scenario where the client's brief asks you to *guarantee* the energy performance of a project. How confident would you be if you were asked to provide that for your current projects? What tools would you need to make the difference? Now think that you may be asked to justify the business case for a new building as part of the strategic definition or briefing stage.

Previously, a client may have said, "I want a new, energy-efficient office building that's going to save our company money." This request is very unspecific. Replace it with: "In proven, measurable terms, I want our new building to save 25 percent on current energy costs and for related improvements across the business to repay the capital costs of its construction."

If this were the case, contracts would need to shift to reflect the new balance of risk and liability. The tradition of unrealistic lowest-tender prices would completely disappear.

Understanding augmented reality (AR)

You've probably seen some kind of demonstration of augmented reality (sometimes abbreviated to AR). *AR* provides you with a live view of the physical world but augments it by layering virtual imagery on top. For years now, AR has seemed like a fun way of overlaying additional information around you; for example, layering history onto your experience of a city tour or receiving entertainment or advertising content through your smartphone camera. A clever recent app was a furniture catalogue by IKEA that acted like an AR target and could make any item of furniture appear in your home via a smartphone or tablet — much better than grabbing a measuring tape (see www.youtube.com/watch?v=vDNzTasuYEw&feature=youtu.be).

For now, you can use AR to view 3D information from 2D targets and explore BIM data and fly-throughs from printed markers. Explaining a set of 2D plans to someone unfamiliar with that way of seeing things is far easier now.

More recently, AR has been in the headlines because of mass-market concept versions of hands-free devices like Google Glass (www.google.com/glass/start) and Meta's Spaceglasses (www.getameta.com). These have begun to demonstrate not just entertainment opportunities but also business, design, and industrial applications, with varying levels of success and adoption.

From an industry point of view, imagine these scenarios using hands-free displays to visualize BIM data:

✔ A site contractor can accurately locate, select, and install correct components based on AR information being delivered to him through a hands-free display.

✔ Site foremen or design teams can show a client the next stage of the construction process when visiting the site, watching time-lapse development of the virtual project in front of their eyes.

✔ A plant operator can expose location models of underground services overlaid on the surface where excavation is required, reducing the risk of cutting or damaging pipes or cables.

✔ An installation engineer can call up instruction manuals or best practice guidance and overlay these to ensure he carries out the work correctly.

✔ A facilities manager has X-ray vision, using the display to augment a view of a wall or floor with model representation of all the services and pipework running through it with their precise locations, flow rates, and maintenance schedules available in the field of vision.

Refer to the nearby sidebar for more examples of AR use.

Viewing information with augmented reality

Wearable technology is guaranteed to grow as more and more devices connect to people's clothes, watches, and glasses, and especially site safety equipment. As we write this, the first iteration of Google Glass hasn't succeeded in revolutionizing the way that people interact with their computers and smartphones, but it has provided enough evidence for innovators to explore other ideas and for other companies to begin to differentiate themselves. Its most public issues were whether it posed a distraction to driving and what privacy issues it opens through hands-free (secretive) recording.

Now, Microsoft has introduced HoloLens, another set of AR glasses using simulated holograms, sound, and motion sensors to simultaneously understand the surrounding environment and project information onto physical surroundings. You'll have the ability to manipulate the imagery with gestures and voice-based instructions; the world around you will look very different. In particular, you and multiple other users will be able to experience the same holographic space, which is immensely powerful for collaboration. The demonstration video shows someone providing instruction for how to fix a sink via Skype, remotely drawing holographic arrows on the pipes and fittings. Alternatively, you could design in virtual space and send your object to a 3D printer (www.microsoft.com/ microsoft-hololens/en-us?video- url=vdeHHP).

Meanwhile, probably the most interesting and secretive application is by a company called Magic Leap (www.magicleap.com) that recently received $542 million in funding from investors like Google and Qualcomm. Its solution creates a "lightfield" of light rays that help retinas to understand the depth and motion of the images being displayed, rather than tricking them with stereoscopic 3D. In Chapter 20, we look at virtual reality and some of the companies in that sphere, but it looks like Magic Leap wants to do something quite amazing with blended reality.

Strengthening Your BIM Foundations

BIM foundations are the solid methods and procurement routes that encourage collaborative working in your organization. In common terms used in this book, we describe these as Level 1, projects with basic digital frameworks, and Level 2, creating a virtual environment of data sharing. (Check out Chapter 5 for more information.)

Think about the future of BIM with that diagram in mind. After a company has built the fundamental essentials to get BIM implemented, then you're about to build the superstructure of an open process based on sharing.

Ensure that you have the strongest possible foundations from which to build. The following sections look at aspects of ideal processes that can strengthen your BIM implementation. Encouraging interoperability, requiring trust among teams, making the most of the power of the cloud, and keeping your data secure will become central to successful work.

Enforcing interoperability

Part of the future of BIM is about encouraging interoperability, if not legislating it. The industry has made great steps forward in collaboration, but it's not very integrated. Standards, like BS 11000, talk about building a joint approach and collaborative working relationships based on mutual benefits. The incentive is that the relationship adds value to the project and encourages long-term working arrangements to maximize these benefits.

Until the regulations and protocols begin to enforce interoperability and increase the requirements for open standards and formats, such as IFC, contracts ideally need to be chosen with collaboration in mind and may be used in tandem with a BIM protocol to ensure the commitment is there from the outset.

In addition to interoperability, other considerations such as capability assessment measure the competency of the supply chain. A number of elements are needed to move toward an integrated BIM, not just a collaborative BIM. The ideal principle is a single, central model rather than a federated collection of individual models. Not only that, but as you start to think about connecting everything in the Internet of Things (see Chapter 20), interoperability is pretty much essential.

In Chapter 10, we help you get your head around collaboration via Construction Operations Building information exchange (COBie) and IFC. It's important to point out that they may not exist in a few years' time, at least not in their current forms. COBie is the lowest denominator to transfer data from you to someone else, because most people have access to basic spreadsheet software. Honestly, though, it's not very sophisticated; these formats are just stepping-stones to real, coordinated solutions that can make the information easier to understand quickly. Open data is vital, and the industry will find new ways to make information interoperable. See Chapter 8 for more information on producing a master information delivery plan (MIDP), which is one of the key components of defining collaborative needs in BIM.

We hope and expect that governments with construction agendas will stimulate interoperability across the supply chain through regulation and new contracts, and promote open standards in other sectors too. Positively, some governments lead the way in opening up their own data for public consumption. Take a look at the United States (www.data.gov), the European Union (http://open-data.europa.eu/en/data), and the UK (http://data.gov.uk).

Trusting people and developing skills

Trust is pretty fundamental to BIM implementation being a success across an organization or across the industry. If you don't trust the other members of a project team you're working with, sharing data can be a risky business. Eventually, you'll work in digital environments where the legal framework is firmly built to defend the rights of designers and authors, but those discussions are still in their infancy for BIM. We cover security, ownership, and legal issues in Chapter 14.

It's probably not going to be easy to get people to share their data with you. Silos will take a long time to disappear. A lot of people will be protective over their data far into the future, especially if that data has provided the commercial advantage that's been their livelihood.

What new skills does your organization (and/or you) require for the future of BIM? In particular, think about how industries like finance, technology, and pharmaceuticals . . . in fact, any big business really needs data, statistics, simulations, and the relevant professionals to interpret the information. We think increasing numbers of businesses will focus on the analysis of the construction industry's data.

Processing with cloud power

You can now access computing processing power in the cloud of a level that once was so expensive that it was out of the reach of mere mortals and the processing power itself was almost unachievable. Cloud computing can increase your ability to work responsively in real time.

Instead of needing to invest in pricey in-house infrastructure, cloud virtualization takes the outlay cost and maintenance overheads of IT off-site. Accessing apps or virtualized software in the cloud, instead of needing processors at the point of use, means that power can all now be remote. Recently, we heard about an office that's gone through this hardware sea change. Instead of spending $3,000 (£1,950) on a new, high-spec laptop, all users can have $300 (£200) laptops and access everything they need in the cloud.

More and more, BIM is going to need the power of the cloud and the huge data centers providing hosted desktop offerings that will lower the cost of handling the federated models you'll soon encounter, which are likely to be significant file sizes.

We're firm believers in the idea that BIM doesn't equal expensive. Some people choose to invest thousands of dollars or pounds in new kits and all kinds of leading-edge software. Vendors everywhere try to convince you that their hardware, tool, or platform is essential and fundamental to doing BIM. We think this simply isn't true.

Of course, some software, programs, and platforms will make your life much, much easier. What's important is looking at your budget and your processes and working out where you can make the biggest efficiencies or time savings. A simple idea like building a master library of BIM objects, specification data, cost information, and technical standards could save many days on a live project.

That computing power will allow for two further shifts in the industry:

- ✔ Extraordinary levels of visualization will allow project teams to simulate and demonstrate the effect of a new building in terms of its environmental, social, or physical impacts and display this in virtual and augmented worlds.

- ✔ Running near-endless algorithms and calculations to optimize design geometry will result in new engineering and construction outputs never previously attempted because of their complexity.

Keeping your digital data secure

Digital data security and cyber security are at the top of many businesses' agendas for investment. As Chapter 14 demonstrates, many of the key standards, like PAS 1192:5, are taking the digital security of BIM very seriously. The UK government has gone as far as to use the phrase "security by default" as a rule for future software/hardware solutions and investment.

It's no longer a question of work and business data moving between discrete technologies; a new age of data security needs sensitive data to move safely between trusted and personal devices as well. More data exists and it's moving faster between more people. Quickly, the construction industry will move closer to the financial sector in how companies will transfer files.

The security of your data, especially what you consider private, personal, or confidential, is going to be subject to increased risk in the Internet of Things. Yes, buildings can generate all this amazing real-time data, but the risk is that this data will leak in places, potentially to people who would use it unlawfully or to do harm. What about malicious attacks on building management systems? Buildings need to become as digitally secure as they are physically. Even the *Mission Impossible*-style "self-destruction" of data won't just be the stuff of spy films but will become an integral part of modern business applications.

Preparing Yourself for the Future of BIM

The future of BIM is going to result in the following:

- ✔ A hyper-productive industry, responding to the need for more and more buildings
- ✔ Model and object intelligence used at every stage of the project
- ✔ Buildings that can positively change life and influence human behavior
- ✔ On-demand information, open and accessible

Great, but don't get ahead of yourself! The industry still has so far to go. You probably agree that construction doesn't handle risk or litigation brilliantly. Too many contracts are designed almost with the assumption that things will go wrong, and people lack motivation to collaborate to get things right.

These sections look at two ways that the future of BIM will develop to use sensors to collect and interpret how people use the built environment and how buildings can begin to influence how people live and work.

Collecting building data and real-time telemetry

One of the key visions for the future of BIM is about using building sensors and real-time monitoring or building telemetry. *Telemetry* just means automatic or remote measurement of something. These sensors can then transmit data to monitoring systems as part of building intelligence and management. For facilities managers and owner/operators, many possibilities exist, such as:

- ✔ Sensors to indicate wear and tear on mechanical parts in elevators, escalators, and moving walkways, to trigger alerts or alarms when servicing is due, and even to predict when other components are likely to fail.

- ✔ Sensors to measure resource use from electricity to water and to clearly understand where demand in a building is highest and when in any given period of time demand peaks and falls.

- ✔ Sensors to alert managers of security breaches, fire, and damage warnings, providing the ability to quickly locate issues on model visualizations.

Nanosecond procurement means that real-time information will likely demolish trading boundaries that are traditionally in place between nations and across borders. It will open up huge global opportunities. (We discuss nanosecond procurement in greater depth in Chapter 20.)

In Chapter 18, we mention that mergers and acquisitions are probably going to form giant superbrands in the construction industry, just as they exist in consumer food, technology, and pharmaceutical spheres. Now consider that these global brands will be able to trade on an unprecedented scale with clients, project teams, manufacturers, and management companies. BIM on just one project isn't really big data, but think about how buildings and infrastructure connected via the Internet of Things are producing huge amounts of data. These construction superbrands will be collecting so much information in the way they understand the use, operation, and performance of the built environment that they'll approach levels you can call big data. You can be sure that clients will clearly monitor and publish carbon cost.

People across the industry will feel the effects of building telemetry:

- ✔ Buildings will be part of sensor-rich networks and connected smart grids.

- ✔ The industry will produce rich, unstructured data.

- ✔ The business model for new projects will become about feedback gathered in-service.

- ✔ Post-occupancy evaluation won't exist; it'll be remote and constant.

- ✔ Clients' focus will be always on the constant analysis of their assets and where they can improve performance.

Influencing and changing behavior

BIM makes the industry more efficient — great. Everyone saves a ton of money in capital expenditure and the basic cost of the built environment — fantastic. The industry lowers the numbers of injuries and its carbon impact, and even becomes more domestically reliant through increased off-site manufacture — amazing. Governments realize that a huge amount of the cost of a built environment asset is in its operation — so any savings made in operation are worth far more than trying to shave off costs in design or construction phases.

But that's not all; here is governments' big dream. Can you change people's lives with the built environment? What if . . .

- ✔ A well-designed hospital actually makes people better more quickly?
- ✔ A well-designed school actually makes students more likely to find jobs?
- ✔ A well-designed digital business district encourages entrepreneurial activity that puts money back into the economy?

That's worth far more than the energy operational savings from a building. The result of changing lives is one less patient-costing healthcare, one less unemployed person needing welfare, and one more business contributing taxes. You can call this behavioral data. Some people call this collection of personal information through the entire life *womb-to-tomb data.* In Chapter 1, we suggest that the acronym BIM can also stand for Behavioral Information Management.

Digitizing an entire nation

Estonia has recently developed a whole system of data collection as part of its digital drive to become e-Estonia, the world's first totally digital nation. Instead of the country being defined by its physical land, the government wants Estonian nationality to be a digital concept. Everyone has ID cards and through interface readers, all connected by a network called the X-road, residents can access entirely digital records. For example:

- ✔ A child is registered for health insurance from the moment of his birth.
- ✔ If a police officer stops you, you can display your car's service history.
- ✔ Your exam results automatically generate your university entry.

Will BIM even exist in the future?

Many people who think about the future of BIM as we do suggest that even Level 3 BIM will serve a purpose only for a relatively short period of time. What's interesting is considering whether Level 4 (integrated, behavioral BIM) will actually just be a common part of a connected global system of information exchange for every kind of procurement from built environment to product manufacturing.

Figure 19-2 shows an extract from the UK Government's Digital Built Britain strategy document for Level 3 BIM and beyond. It starts at the base, with the traditional BIM description of the delivery phase of a built environment project, in the lowest blue pyramid. The pyramid is made up of regular data exchanges in response to key business incentives, building up an increasing dataset of information about that phase of work across a whole portfolio of projects.

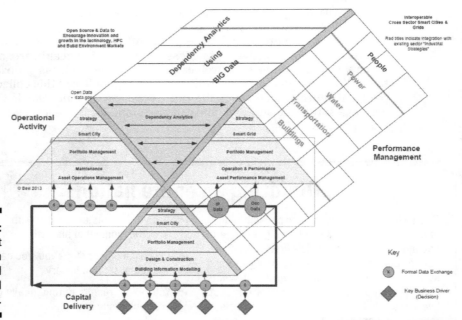

Figure 19-2: Digital Built Britain Operational Model diagram.

Illustration by The BIM Task Group

The assembly of more data increases the strategic importance of information and the ability of an organization (like a government) to use that knowledge for decision-making farther up the pyramid. Above the base pyramid (and connected to it) are similar datasets for the operational and performance management phases of a built asset project. Using big data analysis, you can understand more about the future needs of users.

The long-term vision, shown by the layers of the diagram, is that this kind of analysis and data development extends to transportation, water, power, and eventually entire populations of people. That lifecycle management will help people make much better decisions about infrastructure and the built environment, and by then the industry may not even call it BIM!

Fundamentally, whatever it's called, the process is about information and how it's managed and exchanged. The biggest improvements to the future construction industry will be more open (in terms of collaboration/cooperation) and secure (in terms of cyber risk) ways to host, exchange, engage with, and interrogate data from projects at every stage in their lifecycle.

Noting the political landscape

Governments and public organizations likely will have strong opinions on the future of digital information in construction. Just as occurs with financial and pharmaceutical companies, your global strategy is highly impacted by political attitudes or restrictions. More nations are beginning to require electronic submission of digital information.

Construction strategies are generally pretty safe from government to government. Leaders rubber-stamp anything seen to encourage efficiency and cost "GOOD" (with a digital watermark, of course, not a rubber stamp on paper).

That said, keeping an eye on the changing political landscape is worthwhile to see how the government can influence the future of the industry. If anything, most changes in leadership result in tougher targets and accelerated programs, as an extra incentive to boost the construction economy or to make additional savings.

Chapter 20

Forecasting How New Technology Will Drive BIM

*T*his chapter shows you some of the exciting new technologies that are on the fringes of day-to-day construction, but are already impacting cutting-edge work in the built environment.

Often, future technology is wildly different to what you can see around you today. It's not just innovation that separates past and future, but imagination too. Futurism can quickly become science fiction, but everything in this chapter exists already, even if some ideas are just prototypes or concepts in beta.

You don't have to look decades ahead here; the new tech in this chapter is already on the doorstep and knocking on your door very soon. From a world connected with the Internet of Things to unrecognizable methods of construction and analysis, think of BIM as just the very beginning of the future of an industry ready for change.

Providing a Digital Catalyst

Predicting future technology without being hugely influenced by what's currently around is nearly impossible. The current technology blinkers people's ability to see beyond it. Just consider whether you could imagine Wi-Fi before it was invented. Although flying cars don't exist just yet, plenty of flying information is mobile and everywhere.

Looking to the future

If you've seen the film *Back to the Future Part II*, you know that the character Marty McFly from 1985 ends up 30 years into his future, the year 2015. In those classic scenes, you see an ultramodern cityscape of flying vehicles, automatically lacing shoes, and hoverboards.

Although the makers actually got a lot of 2015 concepts right, like video conferencing, what isn't present is the Internet. In fact, fax machines are considered a high-tech form of communication! That's just one example of how difficult it is to predict the future without being hugely influenced by the present. Today's current technology affects your ability to see

beyond it. *Back to the Future II* is an alternate 1980s, not really a vision of 2015 at all.

If you're interested in hoverboards or how designers can protect future buildings from major environmental events, check out a recent prototype called the Hendo Hover. The board (made by company Arx Pax) uses electromagnetic fields and a copper skate park as a proof of concept for what they call "Magnetic Field Architecture," which proposes being able to magnetically "lift structures out of harm's way," to avoid seismic impacts during earthquakes or damage from flooding. (See www.hendohover.com.)

Think about the concept of email, which took an existing process (postal mail) and digitized it. Now email is just the foundation in a whole range of online communication tools and platforms like Skype and Whatsapp. In the same way, the computer-aided design (CAD) software that the industry has used for a while replicates something paper based in a digital way. CAD platforms and design software are just the foundation layer, and that access to information is the catalyst for the development of innovative ways to work and share project data.

Preventing BIM being just a buzzword

One of the criticisms you may hear about BIM is that it's just a buzzword, or that it's an idea that quickly will be replaced by something else. A good response to this argument is to make an analogy with email. Although lots of new communication tools have evolved from that platform, the basic process of sending digital mail is still used by the majority of businesses and has lasted for a long time since its invention.

By looking at other industries, you can be confident that investing in improving your digital processes is always worthwhile. A balanced approach is getting up to speed with new technology that will take BIM to the next level, so that the appropriate concepts in this chapter are only a stepping-stone away when the time comes for you to make the leap in your business.

Industries that have already gone through digitization are now full of pioneering research and amazing innovation. But the 2015 Ford Trends report found that 64 percent of adults across the globe say technology is moving too fast to keep up with.

Gartner, the IT research and advisory company, produce the Hype Cycles diagram; see www.gartner.com/technology/research/methodologies/hype-cycle.jsp. It describes the way most groundbreaking technology gradually evolves from being an obscure innovation into common usage. Often you see a peak of hype where news media heralds the technology as the solution to everything and then disillusionment when it doesn't quite live up to those expectations, before an eventual adoption when people accept it into mainstream use.

With all new technology, you have a choice whether or not to adopt it in your organization. Traditionally, technology can take a long time to be proven, but increasingly new technologies are moving through the cycle really quickly. If you ignore opportunities now, then the gap between the technology-led expertise and the stuffy, analog processes of old is only going to get bigger, as more and more innovation pushes everything forward. It's always a balance of risk versus reward. The earlier you adopt, the higher the risk you've moved too early. Wait too long and you risk missing out on opportunities.

Here's a good term for the future of industrial activity: Industry 4.0. It describes the fourth industrial revolution, because three previous industrial revolutions have occurred: beginning with mechanization and steam power, evolving with assembly-line manufacturing, and most recently electronics, IT, and robotics making up the third. Industry 4.0 moves automation to another level while also making the most of sensor-rich data and intelligent feedback. In particular, Industry 4.0 is likely to be the first time all industries will connect, from manufacturing to energy, to explore how technologies developed in one industry can be quickly repurposed in another.

Collaborating online

Thinking about online information is pretty magical. When researching this book, we were looking at articles that were probably being read at exactly the same time by someone in India, Russia, and Australia. Perhaps by you! All that content is open to just about everyone with an Internet connection.

Key BIM themes like *online digital collaboration* can seem quite difficult to make sense of. It simply means working together using Internet-based methods of exchange. Users all over the world can join forces, applying trained knowledge and disparate skills to achieve common objectives. This online collaboration sounds a lot like what BIM is trying to achieve. (Refer to Chapter 6 for more information about collaboration and BIM.)

Exploring video games and virtual worlds

Gaming is a multibillion-dollar business and companies put huge investment into gigantic cloud infrastructure (for example, Microsoft Xbox Live runs more than 300,000 servers) just to improve the speed and performance that multiplayer gamers experience. Recently, new innovations in data analysis even match players of similar skill levels or provide fair artificial intelligence for opponents.

Some of the most direct parallels for the built environment industry are the various games in which construction is part of the objective, such as urban planning simulators like SimCity and CitiesXL, and especially the sprawling toy box of Minecraft.

Minecraft is a near-infinite world with various aims, one of which is to build constructions like houses, caves, bunkers, and lighthouses with 3D cubes representing different materials. In Creative mode, you have unlimited resources and can just build whatever you want. How brilliant is it that children (and many adults) are designing and building imaginative constructions, from fantasy castles to pirate galleons? Mojang, the game's developers, even published an official Construction Handbook. That's right — a fun, open-world video game with a construction guide.

Most players are very collaborative, and groups of friends all over the world are learning how to work together, with 3D objects in an open, digital modeling environment. Sound familiar? This sounds a lot like BIM.

In the latest version of SimCity, the city models provide data visualization showing the impact of your decisions on pollution, education, taxes, and the wealth and happiness of inhabitants. Click on a power station and you see its current output, cost per hour, employment levels, the training level of workers, and how safe the plant is. This is *interrogating the model*. By joining multiplayer modes, users can collaborate (and share or sell their resources), including developing self-sufficient regions. Sound familiar? BIM!

There's never been a better chance to include kids in the conversation about a future construction industry. Perhaps you played with Lego or Meccano construction toys. What an extraordinary thought that the games children play in virtual worlds could have the same impact, while encouraging sound engineering, architectural, and construction skills before they've even left school. Not to mention the many who choose to develop virtual worlds themselves. Alison Watson, the founder of the UK's "Class Of Your Own" DEC training, is a great advocate of this; see Appendix C.

For a true connection between BIM and gaming, check out recent apps from Severn Partnership's sister company Seeable (www. severnpartnership.com/services_ item/seeable-apps). The apps use data from 3D point-cloud scans alongside additional visual triggers more commonly seen in video games to make BIM data more accessible and open it up to nontraditional and (nontechnical) users.

You're more than likely already collaborating online. You may share photos with friends and family on social media platforms like Twitter, Facebook, and Instagram, and people probably comment on them or link elsewhere, like sharing data. In academic or office environments, you may do group work on virtual files such as reports developed by multiple authors. Have you ever made a change on Wikipedia or written an article yourself? People make thousands of changes all the time. These aren't perfect analogies for BIM, but they're good examples of cooperative working.

One of the best examples of digital collaboration and perhaps the most exciting is online gaming, especially the massively multiplayer online role-playing games (MMORPGs) like World of Warcraft and EVE Online, and multiplayer games like Call of Duty on Microsoft Xbox Live, Valve's Steam, and PlayStation Network. Refer to the nearby sidebar for more on digital collaboration in video games.

Communicating and sharing data

A big part of BIM is sharing information and using it to make better decisions. One area where future technology is going to impact all industries is more systems that constantly collect data as people live, work, and shop while connected online. Already, these systems are storing unprecedented levels of information.

What's the point in having loads of data if you aren't able to interpret it and gain from the "hidden" information within? Billions of pieces of data may exist that nobody's looking at.

Think about how retail companies collect and use your data. Initially, marketing companies tailor their advertising more closely to your interests so you're more likely to buy their products, but service companies also tailor finance and information products to your personal needs. You can apply exactly the same logic to building information. The data that sensors in future built assets collect will benefit commercial applications and will also improve user experience in new projects.

Without trying to sound like an episode of your child's favorite TV show, *sharing is important*. Increasing access to important data is going to include a bit of risk sharing and a few nondisclosure agreements. Sometimes the data may not be useful to you, but could really help out someone else. Today's episode of *For Dummies* was brought to you by the letters B, I, and M!

Revolutionizing Construction with Big Data

IT companies usually describe *big data* as being sets of information with the following characteristics:

- ✔ Extra-large volume of information (lots of data too big for manual processing)

- ✔ Constant data flow (coming through very fast, too quickly for traditional sorting methods)

- ✔ Varied forms of data (not just a few types, a range of sources providing the information, too complex for one-size-fits-all solutions)

A single building information model isn't really big data. On any given project — even the largest complex buildings, like airports, hospitals, or laboratories — the amount of data being processed is really small for big data specialists, who are used to dealing with billions of bytes (units of digital data) across hundreds or thousands of servers.

However, buildings will begin to use sensors and telemetry to provide constant information to develop information models, and plug into connected, smart cities and grids. Suddenly, across a portfolio of assets or an entire company's output, the costs or energy consumption of multiple buildings become much more like big data. Information collected from many single BIMs will come through thick and fast, in lots of different forms.

You may have heard of the Large Hadron Collider (LHC), a giant physics experiment in Switzerland. It's a particle accelerator, firing super-small particles around a huge circular tunnel underground, and it aims to solve some of the most interesting questions in physics. The LHC shoots *hadrons* (a physics term for one kind of really tiny subatomic particle) around the tunnel in opposite directions, so that they smash into each other. Every second, the facility collides hadrons together 600 million times. That's 600 million times. Every second. The research experiment is collecting insane amounts of data: big data.

For too long, people have never captured valuable information about the use of built assets at this kind of scale and frequency. It's reasonable to anticipate that across the world's 2 billion buildings, sensors will be providing big data to the construction industry. That means the future of BIM requires new techniques and systems to interrogate more data than the construction industry has ever used before. This increased access to data will inform your future projects and make them more efficient, safer, and more cost-effective.

The following sections look at the processes, tools, and companies developing how big data can be applied to the everyday construction industry. We discuss that you may need new job roles to make the most of your built asset data and suggest that project teams can automate many existing manual tasks like compliance checking.

Empowering your business with information

One of the major changes in the last few years is that you don't need to be a statistician to interpret all the numbers. This trend will continue. New software and tools are making it far easier for you to see the power in your data through visualizations. We're convinced that built environment owners, designers, contractors, and operators will begin to connect their BIM data together to understand their portfolios as a whole and see where they can make new savings.

One key development in BIM and knowledge management is the need for new roles in the construction industry. Most likely more industry professionals and built environment companies will employ data scientists for deep analytics. You'll be able to add more value to projects than ever before if you have new insight by using connected data to become indispensable to your clients, co-workers, or customers.

You need to understand what your new information means and apply the lessons to the next day, the next project, or the next building. You can take lessons from the successes and, more so, the failures.

Moore's law is about how many transistors you can fit on an integrated circuit, and roughly that number has doubled every two years — a good general rule for computing power's exponential growth. Recently, Mark Zuckerberg of Facebook described how much information people are putting out into the world and suggested that the amount of data people share will also double every year, just like Moore's law. With all the sensors and devices around you, the amount of information you individually *share* may double every year or two years. In your personal life, you may feel that sharing has become natural to you and you may not even notice how much private information you're sharing. Well, that's happening with building data now too.

Sensing building users' behavior

Big data analysis from the built environment could generate new efficiencies in the way people use buildings and built assets, especially in terms of energy and cost savings. But it relies on knowing a lot more about users than owners and operators ever have before. The future of BIM and big data could also have a negative effect if users are unhappy about sharing data about their activities and movements and if all this data capture opens up buildings to new security risks.

In Chapter 15, we mention how important carbon targets are to governments all over the world and that massive problems exist with how humanity uses existing buildings. The majority of existing building stock is still going to be around in 25, even 50 years' time, so you need to know how it's performing and then make renovation as efficient as possible. Two of the fundamental savings that BIM and big data can generate are

- ✔ **Carbon cost:** Through big data, clients, designers, and operators possibly can understand how different building types, materials, orientations, and energy schemes are performing across multiple locations. Guaranteeing the performance of buildings against anticipated energy usage will be much easier.

- ✔ **Financial cost:** Large contractors and multidisciplinary organizations are already comparing all of their projects around the globe in terms of cost data to understand the true cost of construction. Assessing where savings can be made, either by sourcing materials and products in different locations or by using this information to make tender prices far more accurate, becomes possible.

In the future, understanding how energy use and cost data varies across the built environment as well as how people use buildings differently will be possible. It's not science fiction to imagine individuals being tracked around urban environments, including the interior of buildings, to assess how spaces and functions are being used and where bottlenecks of people occur. This information can then form a feedback loop that can help new designs better meet the needs of users and improve efficiency. You're already being tracked around airports for security purposes, entertainment parks like Disney to optimize sales and marketing, and some conferences to inform the layout and content of future events.

In the 1950s, it became popular to have time and motion analysts in factories, to note where new efficiencies were possible. It started as a good idea. (Every item should be within reaching distance — why should someone walk from her desk to pick up an item and take it to where someone else fitted it?) But then quickly workers resented the analysts (Sorry, your bathroom break was too long). A new way to think of BIM as behavior management is that governments and local organizations could measure positive improvement in social or behavioral activity, such as faster recovery in the hospital or better results in schools. (Chapter 19 discusses behavior management in greater detail.) So where does that take you eventually?

The number of sensors in the built environment will likely increase along with more wearable technology for building users. Some people say all this tracking is all bad news. They see behavioral monitoring, like a black box in your car that could lower your insurance depending on your driving style, as Big Brother–style surveillance. Here are a couple of imaginary scenarios:

- ✔ You've been in six different fast food outlets this week. Your health insurance premium just increased.

✔ You're spending money at a casino. Your personal loan request has been refused.

Whatever sensors and observation you put in your projects, consider the privacy issues that could result. In particular, increasingly privacy-conscious Millennials may resist the idea of sensors hidden everywhere and in everything. BIM in the future will require some ethical decisions about the collection of built environment data and how it's made available for shared use. Here's the general rule: If you're not comfortable sharing information about yourself, don't expect others to be comfortable with your building collecting that data about them. People are generally more prepared to share data about their activity if it's *anonymized* (which means that no personal information is collected that can be used to trace data back to an individual, just generic facts) or if it's clear what the purpose of collecting data is. Secretive collection of data seems more sinister than openly explaining the benefits of information gathering.

Rethinking building: Google and Flux

Although the construction industry has lagged for a while, it was inevitable that some of the big names in technology would eventually look at urban development for new opportunities. The UN Human Settlement Program estimates that 3.5 billion people live in cities, and by 2050 the population will have grown by 3 billion more. So alongside the big BIM platform vendors, you'll see organizations like IBM and Microsoft joining the sector in various ways.

Google recently announced the launch of Project Tango (www.google.com/atap/project-tango), a collaborative effort to develop mobile devices that understand their own motion, along with spatial depth and the layout of objects around them, perceiving how to navigate 3D space and informing users about the 3D movement of devices.

Project Tango could lead to a whole additional layer of big data about urban environments. For example, imagine surveyors being able to use cellphones to quickly map an internal space without the need for expensive laser scanners. In the future, you'll be able to call up building navigation that is the equivalent of using a satellite navigation system indoors. The aim is that the device can generate 3D models of spaces, so as soon as someone maps the space, the plans could become as public as existing ones of museums and galleries. This is building big data, because Google will store the information, where it can be maintained and analyzed as needed.

One of the most interesting innovations has been from Flux. Originally, Flux was part of *Google[x]*, Google's try-anything idea factory. According to their website (https://flux.io), Flux's mission is to "help meet the world's demand for durable, sustainable buildings by applying principles from software engineering, scalable computing, and open platforms."

After various secretive guises and codenames over the years, Flux broke out on its own with some revolutionary concepts. Flux is trying to solve some key problems with the traditional construction industry. Ask yourself the following questions about the future of the built environment:

(continued)

(continued)

✔ **What if you could design building "seeds" that you could let grow according to the site constraints?** You can hear about this in a charming demonstration by Jen Carlile, a co-founder of Flux, at `vimeo.com/107291814 - t=384s`. She explains how the inspiration was Monterey Cypress trees, which, despite having the same basic seeds, adapt in relation to specific environmental conditions and produce different shapes. So in the building seed you have the rules for that particular building type and it can adapt to your project's site conditions, automatically optimize the best design solution, and repeat this over and over.

✔ **What if you didn't need to design a building from scratch every time?** Flux says that current building design is very wasteful both in material and resources. To a programmer, a building is just a solution to rule sets and constraints, so finding an optimum outcome that provides maximum floor area or maximum structural efficiency should be possible with the right algorithms. Imagine if you could program sustainability into every design by selecting the most suitable construction with the least impact and coordinate that decision with everyone on the project team.

✔ **What if the built environment was data-driven, not desire-driven?** Some of the constraints for a project are relatively constant, like a historic zoning code. You can see this theory in their first release: a beta called Flux Metro (`https://flux.io.metro`), initially for Austin, Texas. You can interrogate a plot of land and assess the maximum buildable volume based on planning restrictions for trees, monuments, or views. Manually understanding the conflicts and relationships between different codes can be tricky, but now you can imagine visualizing the combined picture at a glance and in context.

Flux is aiming not just for optimized and replicated buildings, but Google-scale sustainable solutions. It recognizes that the world needs to house 3 billion more people by 2050 and that housing needs to do as little harm to the planet as possible. It points out that even if the world housed 500 people per building, that's still 6.6 million additional buildings in 35 years. So much of the land, outdoor spaces, and interiors of projects end up underused because people crave space but can't fill it.

In Chapter 19, we speak about how future BIM will justify the business case for new projects. Flux is looking at business-case objectives as a driver for design too. We think Flux will likely look to provide lifecycle cost savings and reduce environmental impact by demonstrating optimized efficiency and the health benefits of good design.

Automating rule checking and machine learning

You need to verify that you've matched building code requirements like fire escape, universal accessibility, and thermal design at regular stages in your project, usually for external organizations to assess. By cross-referencing your project against the relevant codes or by running simulations through the information model, you can demonstrate that you've complied with the requirements. You can show that all users can safely exit the building in an emergency or that you've managed heat loss and gain across your project.

This compliance checking isn't just about building codes either. Certain clients and owners have strict and onerous briefing requirements about program and performance. When you add this to the simulation and telemetry from Chapter 19, you can understand the potential in that overall process. We think a huge increase in the automated checking of design and construction properties against constraints like these will occur.

Another technology, *machine learning,* allows you to take this automation one step further. Imagine you could teach a computer to make decisions for itself, instead of having to control it with programming. Well, you can.

Machine learning is a way of building algorithms that learn from real-time data and recognize patterns. Similar to you making a judgment call or a prediction based on what you already know, in simple terms a computer will get better at a task the longer it gets to perform it. Consumer websites or streaming music services use machine learning to understand your likes and dislikes and suggest new titles and appropriate advertising. It's common and many of the online services you use will apply machine learning to understand you better.

For example, generally at Christmas people like to listen to festive music and famous holiday music on repeat. It's vital that the streaming music service can differentiate between this kind of repeated seasonal listening and the new favorites you're prepared to listen to all year, probably because you've skipped the track a few times in January. That's machine learning.

In construction, machine learning is a perfect match for the sensors and building telemetry, as we discuss in Chapter 19. Can your built environment learn to think for itself? We think it can. Already many examples exist of buildings predicting what will happen soon.

Embedding thousands of sensors working together can provide some key benefits. A good example of this is the collaboration between the elevator manufacturer ThyssenKrupp and Microsoft. The sensors in their systems can tell

- ✔ **Exactly where the problem is:** The key benefits of sensors embedded into the working parts of elevators is that when failure does occur, diagnosing the problem doesn't take hours. The sensors tell the story of exactly what's wrong down to the precise part that needs attention. A lot of cars have diagnostic technology, and buildings aren't far behind.

- ✔ **How hot the motors are getting:** Elevators can warn of changes before they become problems, like a motor sending data to warn about irregular temperatures before it overheats. Microsoft provides cloud-based dashboards for all sorts of analytics like this, and the manufacturer says these early-warning systems provide high levels of uptime.

- ✔ **How often the doors have opened:** Another benefit is pre-empting maintenance. Say that the failure point for doors is thought to be 100,000 opening cycles. ThyssenKrupp and Microsoft can program the elevators to recognize when they're approaching this number and to recommend maintenance. However, maintenance staff can add to the model the status of the doors at the point of maintenance, so the algorithm gets adjusted accordingly, perhaps because maintenance was required sooner or actually the doors were fine.

✔ **How often the doors have opened on certain floors and at what times of day:** The previous two benefits are companywide information, but you could also use machine learning at a building level. If the elevators can learn when the arriving and home-time rushes are and on what floors, they can make sure the cabs are already in the right places and moving at the optimum speeds to match users' behavior.

Using AI and BI

Take machine learning to its logical conclusion, and you get *artificial intelligence (AI),* which is simply the intelligence that computers or other machines can demonstrate, and eventually, how closely it can mimic human intelligence, learning, and perception. By combining AI with big data from the built environment, you can analyze and interpret BIM content more quickly, efficiently, and cheaply than human data scientists could ever achieve.

For example, IBM made Watson, a supercomputer. Watson processes information more like a human than any computer in history. By interpreting language naturally and gradually learning, Watson can analyze the messy ocean of data across the Internet like you'd conduct research for a report or an article, carefully selecting the best information and making intelligent assessments.

This kind of AI can begin to make judgments and draw conclusions with far better access to data than you can achieve on your own. Imagine AI optimizing design forms or finding the most efficient structural solutions. AI will increasingly understand plain-language questions like "What is the most sustainable but durable material for this façade?" and "What are the advantages and disadvantages of this site?"

According to *MIT Technology Review,* Watson is about to overtake human understanding of *oncology,* the study of tumors and cancer. When the article says "overtake human understanding," this means in terms of *book-smarts,* because Watson is able to read the latest research in volumes no human could achieve. So how does this change things? Doctors won't need to read every journal in sight. Watson will provide information and diagnosis, allowing the doctor to care and treat the patient, human to human.

Recently, business intelligence (or BI) software has also become very popular. With BI software, the process of retrieving answers to business questions from big pools of data is simpler and more attractive, often allowing for the design of infographics and visual dashboards. Reporting is no longer in boring, dumb spreadsheets, but live and updatable real-world queries and reports. This is important for BIM because it highlights the need for models to be able to answer plain-language questions written by humans, especially nontechnical users and clients.

Decisions in the future of BIM will be made at lightning speed as systems begin to automate the choices you previously needed to make with your project team. Now they're going to become predictive — perhaps completely out of your hands! What decisions does your building, project, or city need to make? Here are some examples:

- ✔ Open another window to regulate temperature.
- ✔ Open another lane for traffic to reduce congestion.
- ✔ Open another hospital to ease healthcare demand.

Connecting Everything Together: The Internet of Things

You have access to lots of disruptive technologies and processes all working in unison. You have access to more information than anyone in construction's long history. The cloud gives you nearly unlimited processing power, and you can access the data you need, whenever you need it, through the Internet of Things. You can link sensors into a gigantic web of information flying wirelessly all over the globe. This is the Internet of Things (IoT) or sometimes the *Internet of Everything*. The future of BIM and big data will form part of the IoT.

The IoT will use online connectivity in the cloud, wireless technology, and smart sensors to link all your electronic devices, from smartphones and smartwatches to your car, home, and workplace. The IoT is predicted to link 200 billion devices by 2020. For example, you can link your utility meter to your home's electronics to control automatically how much energy you're using by turning items on and off, and letting your heating or air-conditioning systems learn from your activity patterns. The IoT could be used to anticipate health problems in at-risk groups by monitoring statistics and vitals and alerting healthcare professionals when emergencies occur. For buildings, the potential of connecting computer systems, alarms, sensors, and building services is vast.

You can use this kind of connectivity in some great ways. For example, a doorbell designed by David Rose at MIT uses location data from smartphones to inform families when everyone has crossed certain thresholds, like 10 miles from the house, 5 miles from the house, and so on. You can see a video of the concept at `https://vimeo.com/40842568#t=112s`.

What's interesting is some of the companies getting involved. Nest, which began making smart thermostats so you could remotely and automatically regulate your heating and smart smoke alarms that inform maintenance teams or emergency services before fire alarms are sounded, is now owned by Google. Nest sees the future not just in the smart homes concept, but also in the information and big data that an organization like Google can gather from these types of sensors. As many others have pointed out before, in the case of thermostats, it's actually in the interest of energy companies to help you use less energy, because so much is wasted. This kind of big data insight can revolutionize the energy industry.

A lot of cynicism and concern still exists about the IoT, just like the behavioral data we mention in the section "Sensing building users' behavior." Not only do some people think connecting sensors and electronics only has value at a huge scale and not domestically, but privacy concerns exist here too.

In particular, are you comfortable with your data being accessible to people who are just trying to sell you more stuff or who may be able to use it unlawfully or to do you (or your building) harm? Not just that, but are people too reliant on technology that can be breached, hacked, disabled, or shut down permanently?

On the other hand, imagine a building that not only pre-empts maintenance issues and regulates energy usage but instantly senses fires and security or digital data breaches in your company. What value would you put on that peace of mind?

The IoT market is set to reach $500 billion in the next five years, and predicted to comprise nearly 30 billion devices. The construction industry isn't just going to be impacted; it's going to be one of the primary marketplaces for the technology, and you need to think about how you can make the most of it.

What role can you play in an evolving new era of construction? What is the new digital you going to be capable of? Can you be the next disruptor of industry? You as an individual have the power to change everything for the better. You can be at the vanguard of the new opportunities. You may just be the innovator who changes construction and built environment industries in amazing, incredible ways we could never have imagined and didn't think to put in this book!

Part VI
The Part of Tens

Read a list of ten great Twitter accounts to follow for BIM discussion at
www.dummies.com/extras/bim.

In this part . . .

- ✔ Discover ten BIM tools and platforms you may want to consider for your business and that will get you on your way with your BIM journey.

- ✔ See how much free information and content is available from supporting organizations, social media, and online groups, and how the online BIM community is a sociable and helpful group of people.

- ✔ Find out the answers to ten key questions that your colleagues, clients, and contacts are very likely to ask you.

- ✔ Cut through the hype and misconceptions surrounding BIM and understand the facts from the fiction.

Chapter 21

Ten Types of BIM Tools, Software, and Platforms to Consider

In This Chapter

▷ Eyeing the different tools available

▷ Identifying the right tools for you

The number of different BIM software and tools on the market today can be bamboozling. We can describe *BIM tools* as task-specific applications that produce specific outcomes. For example, tools are available for drawing production, specification writing, cost estimation, model checking, viewing, and scheduling. Some tools can undertake a variety of tasks; for example, some construction management tools can carry out scheduling, 4D scheduling, and model checking, and other tools may just focus on a specific task.

You can't buy BIM in a box (as we discuss in Chapter 7) because it's a process that's enabled by technology. No one piece of software is likely to fill all your requirements; however, in this chapter we list ten different types of BIM tools that you may want to consider when embarking on your BIM journey.

BIM Authoring Tools/Platform

BIM authoring tools (or *BIM platforms,* as they're sometimes called) are usually used in the design for generating data for multiple uses. Although generic authoring tools exist that cover the basics of architecture — structural and MEP, for example — some tools are specific to a discipline. *Parametric authoring tools* use a combination of graphics and information to allow a design to develop from concept to construction documentation. Vendors such as Autodesk, Graphisoft, Bentley, and Tekla are the main players in this sector. Each piece of software works in a similar way, but all have their own idiosyncrasies.

Some authoring tools focus on feasibly and preliminary design and may not offer interoperability capability, such as the import and export of Industry Foundation Classes (IFC) information.

BIM Analysis Tools

With the project team creating all this digital data and information with BIM, its real value is in how you make sense of it and understand what it means or is telling you. BIM analysis tools enable you to perform a variety of analyses in the design and development of an asset. You may require different types of analysis at different times during the lifecycle.

At an early stage an energy model gives you answers to issues around orientation, form, and occupancy levels, and aids in calculating the asset's energy requirements. Other analysis aspects may include things such as thermal performance, lighting levels, wind flow, and pedestrian flow. Even though some analysis tools are standalone versions, some are add-ons or plugin modules for other pieces of software.

Specification Tools

Not all information is modeled graphically. Within the coordinated data environment there will be graphical information, nongraphical information, and associated documentation. Specification tools play an important part in the nongraphical information, which evolves over the project timeline. In Chapter 10, we discuss some instances where it's more appropriate to use the written word because it can take you to a deeper level of information.

More sophisticated specification tools allow the project team to coordinate the specification and the design model by using plugins. A *plugin* (or *add-on*) is a software component that adds a specific feature to leading BIM authoring tools. Some specification tools also allow the project team to publish information in a variety of different views, reports, and file formats, such as XML, for data schemas like COBie.

File-Sharing and Collaboration Tools

Think of file-sharing and collaboration tools as the common data environment (CDE) (as we discuss in Chapter 4) that provides document management. These tools are at the heart of the BIM environment. Many of these applications run in the cloud and as platform as a service (PaaS), which we discuss in Chapter 7. They allow you to work from your desktop, the Internet, or from a mobile device.

The project team shares all types of information via these tools, including contracts, schedules, specifications, reports, and model information. They vary in sophistication and functionality, from simple uploading and downloading of information with audit trial capability for record keeping and access-security control, to inbuilt model viewers and instant messenger systems.

Construction Management Tools

Construction management tools allow you to take an overview or holistic review of model information. Historically, there was a disconnect between CAD and project management software, with CAD software concentrating on geometrical graphical information and project management software focusing on scheduling but with no integration between the two. However, the line between the abilities of the two is becoming blurred. By integrating the 3D model with popular project schedule applications, you get a 4D model showing construction sequencing, simulations, and scheduling capabilities.

As Chapter 17 discusses, the ability to see the construction sequence and plan for on-site clash detection can be extremely beneficial when it comes to health and safety and the prevention of accidents and injury.

Model Viewers and Checkers

Although some collaboration and construction management tools have inbuilt model viewers, a number of dedicated model viewers are available and a large proportion are free. These usually allow functionality such as combining and viewing models. Model viewers use open data standards such as IFC, which we discuss in Chapter 11; however, some software runs from property file formats.

Premium paid-for versions offer the capability of the viewing tools but have added functionally such as model checking for clash-detection purposes based on rule-based checking for compliance and validation of all objects in the model.

The BIM Task Group lists some free model-viewing software. Check out www. bimtaskgroup.org/free-bim-vewing-tools.

Quantity Takeoff and Estimating Tools

A number of approaches to BIM-based quantity takeoff and estimating exist. Although many BIM authoring platforms can perform an automated quantification of items, areas, surfaces, and volumes of the asset and export this information to a spreadsheet format, they don't produce a cost estimate.

Tools used for quantity takeoff and estimating usually link to BIM authoring tools via plug-ins or embedded BIM information from a BIM authoring tool into a quantity takeoff tool to automatically extract information and necessary quantities from 3D geometric data.

Three-dimensional objects within a model contain geometric information, such as volumes and areas, which can be analyzed. Usually, quantity takeoff updates automatically when the project team updates the base model information. This information then informs cost estimation. It's important to remember that an accurate quantity takeoff requires an accurately built design model in the first place.

Shop Drawing and Fabrication Tools

As you discover in Chapter 10, the level of detail and level of information within BIM objects are usually just enough to imply a product rather than manufacture it. In order to take that information through to shop drawings and fabrication, you need an additional level of information. The project team can transfer information from BIM authoring tools into fabrication software where additional detailing can commence.

Facility Management

Computer-aided facilities management (CAFM) can comprise a number of technologies and may combine CAD systems and relational database software focused on facilities management as well as interfacing with other systems — for example, with computerized maintenance management systems (CMMS).

Similar to file-sharing and collaboration tools, CAFM tools are usually web based. Data is stored, retrieved, and analyzed from a single source. The project team can transfer information from BIM authoring tools into a CAFM system where facility managers can add additional asset data.

Administration Tools

Don't overlook the humble portable document format (PDF), spreadsheets, and word-processing tools. Although most computers come supplied with a basic suite of office applications, they may have limited functionality. If you use a tablet or smartphone while on the move, you may want to consider premium over free read-only versions of office tools, which allow for editing, commenting, and saving.

Chapter 22

Ten of the Best BIM Resources

*T*he Internet has provided a platform for instant access to information and a place in which everyone can come together and share best practice and knowledge. What's more, often this information doesn't cost a dime! However, to really harness its power, the transfer of information often involves a two-way process. A tweet, a post, an article, a presentation, or a good old-fashioned face-to-face conversation imparts to others your wisdom, knowledge, and experiences of hitting your head against a brick wall. As the saying goes, "You only get back what you put in." So if you have something to say or advice for others, share it! Along with the sound investment you've made in the purchase of this book (thank you, by the way), this chapter lists ten great further resources to help get you on your way and how you can capitalize on them.

Using Social Media

As society progresses in this digital age, communication and interaction via social networks such as Twitter, Facebook, and LinkedIn are on the increase. Through the power of tweets, status updates, and links to articles, you can tap into a world of online communities that have a real sense of passion and enthusiasm and an environment of collaboration and knowledge sharing. Individuals should be able to access this information easily and influence a team without having to focus on practice strategies or needing to force their ideas through differing levels of management.

Whether it's a link to a document, a discussion on best practice, or a tweet, the online community provides a vast, untapped, free resource of information and comradeship. This ideology has been helped by various leading industry figures showcasing their knowledge and sharing it with those around them. Twitter and other social media outlets break down communication barriers between small and large organizations, regional or global organizations, and sole practitioners, academics, and journalists.

Twitter is largely self-regulated, and people are free to express their opinions and minds. Just be mindful, however, that you're broadcasting to the world. A general rule: if you wouldn't want your granny to hear it, don't tweet it. Remember that anyone can join in on your conversation. If you tweet something controversial or incorrect, then you're going to know about it.

Don't know where to start or whom to follow? Have a look at *Construction Manager*'s most influential, interesting, and informative Building Information Modeling (BIM) tweeters via #BIMtwitter50 (https://twitter.com/hashtag/BIMtwitter50). Whether a BIM manager, contractor, product manufacturer, architect, or government official, the campaign aims to list the top 50 Twitter users you should be following to be involved in the BIM conversation. Also check out *Building Design* magazine's yearly top BIM tweeters list via #BDTwitter100. The #UKBIMCrew (https://twitter.com/hashtag/ukbimcrew) has become a legacy to the early adopters in the UK seeking out and sharing knowledge among each other on an open platform. Its influence has now moved to #GlobalBIMCrew (https://twitter.com/hashtag/globalbimcrew). The groups aren't exclusive, and nor do they focus on one area of BIM implementation or process. Rather, they provide a search mechanism for the industry to seek out an exhaustive array of discussions, debates, questions, and information all relating to the UK and global BIM communities.

Watching Videos

You may be surprised by how much you can take away from a video. Whether it's fixing a fence or figuring out how to sew a button onto your shirt, sites such as YouTube (www.youtube.com) are great places to gather new skills. YouTube and similar sites are more useful than just for watching videos of cute animals or people falling off skateboards. They're a fantastic resource full of instructional videos and tutorials by likeminded and enthusiastic individuals and companies. A simple search online can uncover many hours of content.

In addition to YouTube, we suggest that you try these sites:

- ✔ **The Revit Kid:** At http://therevitkid.blogspot.co.uk, you can find tips, tricks, and product information created and maintained by the Revit Kid, also known as Jeff. The site was initially started to help his peers. With a limited amount of video-style tutorials available when Jeff was learning Revit, he decided to address the issue head on.

- ✔ **National Building Specification (NBS):** To see what the rest of the industry is up to, we suggest a quick perusal of this site (www.theNBS.com/ BIM) that covers a broad range of programs, including case studies, technical guidance, and interviews with leading industry figures.

- ✔ **The B1M:** Called the B1M because it hopes to inspire a million people through its videos, mobilizing BIM adoption around the world, this great resource (www.theb1m.com/about.asp) has lots of five-minute short videos, great when you don't have a lot of time to spare and want to get straight to the point.

Attending Trade Shows and Events

For those people who stay steadfast in traditional means of communication, events and gatherings are a superb place to not only network with your peers but also come together and share ideas. Although online communication is great, you can't beat good old face-to-face meetings to meet others involved in the field and share knowledge.

Networking at trade shows and events is hugely important, allowing those interested in BIM to meet regularly to discuss themes and the progression of BIM in the industry. Events are organized by professional institutions, software vendors, and like-minded people who want to share best practice. You can also listen to leading industry figures right through to BIM newbies wanting to share their experience through open-mic sessions.

Reading Blog Posts

Many great bloggers are out there spreading the word of the BIM gospel, and a quick Internet search can point you in the right direction. Some blogs are summaries of various BIM events and presentations attended, and others are full of opinion pieces, tips, tricks, and the occasional anecdotes.

Listing all blogs would be impossible, so we offer a few suggestions for you to start:

- ✔ **Construction Code:** Independently run by Stephen Hamil, PhD (director of Design and Innovation at NBS), the site (www.constructioncode. blogspot.co.uk) is a useful resource for up-and-coming information relating to BIM, construction, and technology. He also provides great summaries and round-ups of the various conferences that he speaks at and attends. You can also follow him on Twitter at `https://twitter.com/StephenHamilNBS`.

✔ **Practical BIM:** From Aussie architect Antony McPhee, this blog perhaps isn't as updated as frequently as others, but the site (`www.practicalbim.blogspot.co.uk`) is about quality rather than quantity. Antony provides no-nonsense guidance, opinion pieces, and food for thought.

✔ **The Case for BIM by Casey Rutland:** Check out (`http://caseyrutland.com`). Aside from having the greatest pun-named website, Casey is an architect and associate director at Arup Associates and one of the founding forefathers of the #UKBIMCrew and now the extended #GlobalBIMCrew. He also compiles the yearly #BIMTransferSeason, also known as "The List," that's published as an image on various social media platforms on New Year's Eve. Generally, The List is interesting on a personal level, but it's also the beginning of tracking the head-hunting trends of large companies over the recent years. The List isn't exhaustive and is mainly limited to those who share BIM-related knowledge on Twitter, but it's representative of the demand for BIM experience and expertise in the industry. You can also follow Casey on Twitter at `https://twitter.com/CaseyRutland`.

✔ **Bond Bryan BIM blog:** An award-winning blog from UK architectural practice Bond Bryan, the site (`www.bimblog.bondbryan.com`) is run, written, and maintained by architect and associate director Rob Jackson. The site is full of opinion pieces, top tips, and all things OpenBIM. The site also features downloadable useful documents that cover BIM acronyms, a BIM dictionary, and the different file formats. This site is a great example of a small and medium enterprise (SME) company proving that size isn't a barrier to BIM. You can also follow Rob on twitter at `https://twitter.com/bondbryanBIM`.

Perusing Publications and Journals

Magazines and trade journals are excellent sources of information. Many are now in both print and digital forms, and they also feature regularly updated websites. Although some carry a small subscription charge, many offer a free digital version. Here are two of our top choices:

✔ *AEC Magazine:* Published bi-monthly, this magazine (`http://aecmag.com`) has a UK emphasis and is a good source of news on software and hardware technologies and collaborative working practices.

✔ **AECbytes Magazine:** This online publication (www.aecbytes.com) has been going since 2003. Its main focus is on research and analyzing and reviewing technology products, and it's full of articles including conferences and show coverage, product reviews, case studies, and the usual tips and tricks sections.

Tapping into the Government

A number of national governments have mandated or are intending to mandate BIM. Many include dedicated websites that feature resources and downloadable documentation. In the UK, the BIM Task Group (www.bimtaskgroup.org) provides support, access to all BIM4 Groups, and lessons-learnt documents. It highlights the latest news and gives quick industry updates as well as links to the government projects. The US General Services Administration (www.gsa.gov/portal/category/21062) features a number of downloadable guides as well as details of BIM Champions throughout the different regions.

Signing Up for Software User Groups

Most software vendors support user groups both regionally and internationally. Although some groups are organized and arranged by the vendors themselves, others are established by enthusiastic individuals. Groups are usually free to attend and offer a good opportunity to network and share experiences, as well as a mechanism for feeding back issues directly to the software vendors.

You can find details of many of these groups by searching online or by having a LinkedIn or Twitter presence. Your local software vendor reseller usually can point you in the right direction.

Joining Associations and Forums

Associations and forums are places where people come together (either physically or virtually) to share ideas, opinions, and views on particular issues often through subcommittees or specialist group interest areas. They include the following:

✔ **BIM Forum:** This forum's mission (https://bimforum.org/) is to facilitate and accelerate the adoption of BIM within the AEC industry. The BIM Forum shares user experiences and goals via online forums and practical industry conferences. Sub-forums look at particular issues and topics such as technology, insurance, and academic and legal issues.

- ✔ **Fiatech:** This membership organization (`www.fiatech.org`) considers itself an international community of passionate stakeholders who work together. The group's aim is to lead global development and adoption of innovative practices and technologies to realize the highest business value throughout the lifecycle of capital assets.

- ✔ **Institute for BIM in Canada:** Its mission is to lead and facilitate the coordinated use of BIM in the design, construction, and management of the Canadian built environment. Check out `www.ibc-bim.ca`.

- ✔ **BIM-MEP AUS:** This global leading industry initiative (`www.bimmepaus.com.au/home.html`) strives to address some of the barriers to the adoption of BIM in the Australian construction industry. The group looks to address issues such as standards, practices, guidelines, and work flows.

- ✔ **UK BIM4 Community Groups:** A number of BIM4 groups have been set up to form specialist interest groups who'll champion BIM within their respective specialist areas. Their aim is to develop consistency of messaging in a clear and concise manner to support both new and existing professionals in their respective BIM journeys. They cover a wide range of areas from water and rail to retail and infrastructure, and the future-gazing group, BIM2050. Why not get involved — or, better still, why not set up your own group?

Visiting Virtual BIM Libraries

Sometimes the out-of-the-box content just doesn't cut it or simply isn't available. BIM object libraries can save you time by providing you with pre-configured objects.

Do be careful, because quality and reputation vary from library provider to library provider. Some libraries are community based, so you don't always know the quality of the object you're bringing into your model, and others may only be limited to a particular software vendor's platform.

The BIM object library market sector is a competitive one, and a simple Internet search can bring back a good research list. With BIM being mandated in the UK, a number of UK and European providers dominate within this market, most of which are focused on a COBie deliverable.

Heading to Summer School

Although not cheap, many software providers and software user associations organize a summer school or the strangely named *university*. These schools usually consist of conference training, networking, and sometimes the opportunity for certification, and they often take place over a few days or a week. Although some of these schools go under the guise of a university, actual academic universities and colleges do also run a number of courses.

As well as advertisements in trade magazines or software vender newsletters, check out your local university or college website because you may have a course running on your doorstep.

Chapter 23

Ten (or So) Myth-Busting Questions and Answers

In This Chapter

▶ Looking at cost

▶ Considering how much detail you actually need

▶ Understanding who really benefits

A lot of misconceptions and hype exist around the subject of BIM, with people often overinflating their understanding and knowledge. In this chapter, we cut through the BIM hogwash and dispel ten or so common myths.

Isn't BIM Just Trendy CAD?

BIM is a process that's enabled by technology. You can't buy it in a box, it's not a software solution, and it's something that you can't do in isolation. Although 3D CAD provides huge benefits over traditional 2D CAD, to fully consider the real benefits of BIM you should coordinate the 3D model with information from other areas, such as the master specification.

The *information model* is actually a combination of several things and not just 3D geometry. Information models such as the project information model (PIM) and asset information model (AIM) (refer to Chapter 8) contain graphical information such as geometry; however, these models also contain non-graphical data such as performance properties as well as links to associated documentation such as specifications. The shift from CAD to BIM fundamentally meant creating models rather than drawings and working in a collaborative way.

Is BIM Just for Big Companies, Skyscrapers, and Government Projects?

BIM isn't just the preserve of the big. Quite the contrary. Smaller organizations have perhaps the most to benefit from the efficiency gains that BIM brings, meaning that they can punch well above their weight or bat above their average. Small and large organizations both face the same financial costs, albeit on different scales. Unfortunately, many publications on BIM show a skyscraper on the front cover when BIM is equally applicable to small-scale projects.

Small doesn't always mean simple. The size of the project has no bearing on the complexity of the build, and a large industrial building could be simpler to design and construct than a small retail unit in the city center. And so this brings us to the final part of this particular myth: BIM is only for government projects. Although governments such as those in the UK are mandating BIM more and more, private sector clients and contractors are realizing the benefits that it brings and are also mandating BIM on their projects.

Is BIM a Fad and a Recent Technology Development?

The concept of coordinated design to eliminate waste and increase cost savings certainly isn't new. In the UK, costs savings have been based on research by the UK Building Research Establishment (BRE) in the '70s and '80s. Even the potential of automated code checking for clash detection began as far back as 1966, when SJ Fenves discussed tabular decision logic for structural design in the *Journal of the Structural Division*.

There has been some debate as to where the term *Building Information Model* actually came from. Some say that it was Charles Eastman at Georgia Tech who first coined the term, Autodesk's Phil Bernstein who first used the term *building information modeling,* and Jerry Laiserin who then populated the term. The first implementation of BIM software is often attributed to the virtual building concept by Graphisoft's ArchiCAD in 1987.

Isn't BIM Expensive?

Yes, you may incur more upfront costs when implementing BIM compared to other traditional processes; however, that said the longer-term benefits and rewards counteract any loss. Although BIM isn't technology, software, hardware, and IT infrastructure play important roles, as too does training associated with implementing new technology, which also comes at a price.

When thinking about selecting your digital tools of the trade, think of them as a work horse. The cost of BIM doesn't have to be in dollars and cents. As we explore in Chapter 21, many software solutions and tools are available for free, as are sources of help, training, and support.

Does BIM Adoption Affect Productivity?

Adopting new ways of working is going to take a little time, which is no different to when construction professionals decided to put down the pen and pick up a mouse and embrace computers and CAD. BIM isn't easy, but if you stick with it, you'll become more productive in the long run.

When adopting BIM and new ways of working, make sure you give yourself enough time and take one step at a time. Trying to create a coordinated 3D model for the first time on a Friday afternoon when you need to get information to the contractor perhaps isn't the ideal solution. Can you imagine swapping your computer for a typewriter? No, neither can we.

How Detailed Does the Geometry Need to Be?

Getting carried away modeling every nut and bolt is easy. Generally, the construction industry uses BIM to imply a product and not manufacture it, and excessive geometry just slows down the model. That's not to say that it can't support manufacture and fabrication, just be sure what level of information the client has asked for. You need just enough information so that the project team can specify and choose the construction products, and the level of that information increases during the project lifecycle. For example, at a concept stage you can show objects by way of a *bounding box,* which is essentially a 3D rectangle that represents the object as a stage when its shape and detail have yet to be determined. The box can also show any space or clearance requirements. Consider a chair. (Most CAD standards and manuals like to show a chair for some reason or another.) At an early stage the design team requires just its rough dimensions. As you progress to a design stage, the client may not know what type of chair he wants, so you don't spend time modeling arms, wheels, and other procrastinating details.

The NBS Digital toolkit defines the level of detail required as a minimum for some 6,000 objects at different stages of a project lifecycle.

Do Clients Know What to Do with All the Data?

Often clients ask for their project to be in BIM without really knowing what it is or the benefits that it can bring. Initiatives such as soft landings (which we refer to in Chapter 9) make sure that the client and facilities management teams are brought in early in the process to fully understand their requirements. Soft landings also makes sure that the project team understands what data the client requires and how he's going to use it.

Unfortunately, you don't have the luxury of a crystal ball. However, a client with open-source, structured data can use this in other processes and software, or use the data at a later date if and when the client makes any alterations or modifications.

Do the Design and Construction Teams Only Benefit BIM?

The client can take advantages of BIM by exploiting the information in the model during the design phase to optimize the best possible asset. The cost of designing and constructing an asset pales into insignificance compared to the operational costs over the asset's lifetime. So being able to see where operational costs lie through consistent, structured digital-asset data in order to make post-occupational decisions is a massive benefit. Of course not all clients are interested in the long-term costs of a building; perhaps developer clients are more concerned with letting or selling an asset after it's built.

Measuring BIM benefits and costs consistently and apportioning who gets what benefit and why is difficult. The reality is that everyone in the supply chain benefits when using BIM in a collaborative way.

Do You Need to Push the Detailed Design Forward in the Program?

Today, clients aren't only procuring a physical asset, but they're also procuring information, typically in a digital format. The principles of design development haven't fundamentally changed to harness many of the benefits of BIM. However, a far greater need exists for common, aligned geometric and information outputs. For example, without this coordination the project team and client simply can't achieve quantification, energy analysis, and many of the other BIM uses. However, just because you can model a window and detail hinges and sills from day one doesn't mean that you do; rather, the amount and level of detail increases as you progress through the project lifecycle in the same way it always has.

Appendix A

Glossary

3D model: Model in three dimensions.

4D model: Four-dimensional model (includes time schedule data).

5D model: Five-dimensional model (including time and cost data).

assembly: A physical aggregation of components.

asset: Item, thing, or entity that has potential or actual value to an organization.

asset information model (AIM): Model containing data and information to help an organization manage its asset.

asset information requirements (AIR): Defines the information that's required for an asset information model.

attribute: Piece of data that defines a property or characteristic of an object, element, or file.

big data: Data that's so large it's difficult to handle using traditional database and software techniques and processing capacity.

BIM champion: Person responsible for encouraging and supporting others in BIM adoption and implementation.

BIM collaboration format (BCF): An open XML file format that helps support workflow communication in a BIM process.

BIM consultant: A specific role that may focus on a particular area and works with the organization to support the successful adoption and implementation of BIM.

BIM coordinators: They lead their discipline within the project and coordinate any clash detection and any actions or outcomes that may be required.

BIM environment: Data management that integrates applications such as BIM tools and BIM platforms within an organization.

BIM execution plan (BEP): Pre- and post-contract document prepared by suppliers setting out a structured, consistent process for how the project will be carried out, including common terminology for job titles, descriptions, responsibilities, and processes.

BIM manager: Specific role that is responsible for looking after IT, developing company policies, processes, protocols, and change management.

BIM maturity level: Level within a maturity model that categorises types of technical and collaborative working to enable a concise description and understanding of the processes, tools, and techniques to be used.

BIM maturity model: Illustrates BIM competence levels expected and the supported standards and guidance notes, their relationship to each other, and how they can be applied to projects and contracts in industry.

bimify: The act of retrospectively applying a BIM process to an existing project.

BIM object: A digital representation and placeholder for graphical and non-graphical information about a real-life construction product.

BIM platform: Application that's usually used for design in generating data for multiple uses, such as Autodesk Revit, Bentley Microstation, Graphisoft ArchiCAD, Nemetschek Vectorworks, and Tekla Structures.

BIM process: Process reliant on information generated by a BIM tool for tasks such as analysis, cost estimation, and fabrication and construction.

BIM protocol: Supplementary document incorporated into professional services appointments and used as a legal framework to encourage and promote the use of BIM.

BIM tool: Task-specific application that produces a specific outcome; for example, tools used for drawing production, specification writing, cost estimation, model checking, viewing, and scheduling.

BIM wash: The act of deliberately or intentionally overinflating a person's or organization's BIM ability.

Building Information Model (BIM): A data-rich shared collaborative model containing information made up of many data sources that's maintained throughout the life of a building, from inception to recycling.

Building Information Modelling (BIM) Task Group: A task group supporting and helping to deliver the objectives of the UK Government Construction Strategy and the requirement to strengthen the public sector's capability in BIM implementation.

buildingSMART Alliance (BSA): Leading international nonprofit organization devoted to identifying and delivering construction shared structured data standards such as IFC and ISO standards. Formerly known as the International Alliance for Interoperability (IAI).

buildingSMART Standard for Processes: Specifies when certain types of information are required during the construction of a project or the operation of a built asset (formerly known as the Information Delivery Manual or IDM). See buildingSMART Alliance (BSA).

CAPEX: Capital expenditure.

clash detection: Process of identifying conflicts and issues through 3D collaboration and coordination. Also known as _interference checking_.

clash rendition (CR): The interpretation of the native format model file that's used specifically for spatial coordination processes and to achieve clash avoidance or to be used for clash detection.

cloud: Data and software programs accessed over an Internet connection rather than a hard-drive.

computer-aided design (CAD): Computer technology to assist in the production of design information.

computer-aided facilities management (CAFM): Computer technology to support facility management.

common data environment (CDE:) A single source of information, such as a project extranet where project information is collected and shared among the project team.

concurrent engineering: Methodology of working on a task simultaneously rather than consecutively.

Construction Operations Building information exchange (COBie): Nonproprietary spreadsheet data format that contains digital information about a building in as complete and as useful a form as possible.

data: Information that hasn't yet been interpreted.

data drop: A collection of data exported from the BIM at a particular stage in its history, in a particular format, and in a level of definition.

data dictionary: Collection of terms that provides a language-independent information model that can be used for the development of dictionaries to store construction information.

deliverable: A tangible or intangible quantifiable outcome that's provided to a receiving party.

design intent model: Initial version of the project information model (PIM) showing the designers' intentions.

Digital-built Britain: UK Level 3 strategic plan.

digital plan of work (dPoW): A generic schedule of project phases, including roles, responsibilities, assets, and attributes, that's made available in a computable form.

digital security: The protection and safeguarding of digital data, identity assets, and technology.

designing for manufacture and assembly (DfMA): Consideration of the ease of manufacturing and assembly processes when developing project work.

disruptive technology: Technology that has the potential to dramatically affect the construction industry by upsetting traditional processes and wiping out some obsolescent systems altogether.

Electronic Document Management System (EDMS): A system for the storing, retrieving, sharing, and managing of electronic documents.

element: Digital representation of an object, system, or material, including associated data, for incorporation into a BIM.

employers' information requirements (EIR): Document that defines and communicates the information required by the clients at different project stages.

exchange schema: Method of data exchange for possible mapping to different formats such as XML, databases, and text files. Examples include ifcXML, gbXML, DWG, ODBC, and IFC exchange schemas.

extensible markup language (XML): File format used for the design and structure of documents and data in support of computer-aided BIM applications.

extranet: A private computer network, using the Internet to share and exchange information securely.

federated model: A model consisting of linked individual models and other information to create a single model of the asset.

geographical information systems (GIS): A system used to visualize, analyze, capture, and store spatial and geographical data.

iBIM: An integrated BIM as opposed to a collaborative BIM.

infobesity: The act of putting too much information into the geometric model instead of thinking about relational databases and linking models to other data sources.

International Framework for Dictionaries: (IFD): Reference library to support and improved interoperability between construction information databases and IFC.

Industry Foundation Classes (IFC): A neutral, nonproprietary data schema developed by buildingSMART to define, describe, exchange, and share information.

information manager: Specific role that's responsible for setting up and managing the CDE and managing the exchange of information between parties.

integrated project delivery (IPD): Unites people, systems, business structures, and practices from the commencement of a project to deliver a better outcome.

International Alliance of Interoperability (IAI): Founded in 2001 (renamed buildingSMART in 2007) and originally formed as a collection of international chapters. The IAI was collectively responsible for the buildingSMART concept and the ISO standard of Industry Foundation Classes (IFC) for Building Information Models.

Internet of Things (IoT): The concept of everyday physical objects that are connected to the Internet and have the ability to identify themselves to other devices.

interoperability: The collaboration, exchange, and ability to operate on building model data between different BIM platforms.

level of definition: The collective term used to describe level of detail and level of information. Also referred to as *level of development.*

level of detail (LOD): The level of graphical data at a particular stage of work.

level of information (LOI): The level of nongraphical data at a particular stage of work.

library: Collection of reusable predefined assets such as objects, materials, and textures that you can import into a BIM platform.

master information delivery plan (MIDP): Document used to manage the delivery of information during a project.

maturity level: Establishes at a headline level the baseline levels of implementation, providing clear competence levels that are expected, together with the supporting standards and their relationships to each other and how they can be applied to projects and contracts in the industry.

maturity model: Establishes a datum that the project team can measure against and helps organizations understand where they are now and where they need to be.

metadata: Data that is used to describe other data.

Model View Definition (MVD): A subset of the IFC schema that you need to satisfy one or many exchange requirements of the architecture, engineering, and construction (AEC) industry.

OPEX: Operating expenditures.

Open BIM: Universal approach to the collaborative design, realization, and operation of buildings based on open standards and workflows. Open BIM is an initiative of buildingSMART and several leading software vendors using the open buildingSMART data model.

organization information requirements (OIR): Data and information that is required to achieve the organization's objectives.

parametric modeling: Design using rule-based relationships between intelligent objects that enable the project team to update related properties when one property changes.

plan of works: A shared generic framework for the key project stages where the project team can assign roles and responsibilities; see also *digital plan of work (dPoW)*.

plain language questions: Clear questions that the client asks the supply chain during key stages of a project.

project implementation plan (PIP): Document describing the supply chain's capability and ability to deliver the project, including IT and human resources.

project information model (PIM): A data-rich information model that's developed during the design and construction phases.

property: Characteristics that are assigned to objects to reflect specific information, such as technical data. The terms *parameters* and *attributes* are also used in BIM platforms and construction information. Collectively, these bits of information form a *property set*.

radio frequency identification (RFID): Information that's transferred wirelessly for the tracking of items.

responsibility matrix (RM): A schedule describing who's responsible for specific tasks and deliverables.

schema: A description setting out a document structure so that the data contained in it can interact with similar data in another. An XML schema sets out the structure and contents of a document written in XML.

soft landings: A building completion protocol that involves gradual handover and encourages the greater involvement of designers and contractors together with building users and operators before, during, and after handover.

task information delivery plan (TIDP): Document prepared by individual teams containing a federated list of information deliverables for each task, including format, date, and responsibilities.

virtual design and construction (VDC): Simulation of the construction process using computerized models and techniques.

Appendix B

Your Handy BIM Checklist

So you're thinking about undertaking your first BIM project. If so, this appendix helps make sure your pre-flight checks are ready to go and that you're prepared for a successful and planned take-off.

✔ **Identify what information is required to enable decisions at each work stage.** If your client hasn't defined her needs then suggest helping her with a discovery workshop.

✔ **Plan your strategy with a BIM kick-off meeting.** Ensure you engage the proposed project team and your IT department among others to ensure buy-in. In your strategy, make sure that you

- Identify BIM project goals, both overall and at each stage.

- Identify any contractual BIM deliverables. Can you deliver these in-house or do you need external support such as a BIM consultant?

- Agree key workflows agreed such as clash detection and 4D simulation.

- Agree on a digital project technology stack. A *technology stack* is a set of software that provides the infrastructure for your computers. Check license agreements and hardware requirements.

- Determine the project infrastructure. Ensure you have sufficient bandwidth to push and pull all this data around without long time lags.

- Establish the common data environment (CDE). This is mission critical! Plan this so you can effectively exchange information with all the project participants. We know that local hard-drives and FTP sites seem easy, but they don't allow coordinated collaborative working. Spending time getting naming conventions and folder structures correct is worthwhile; trust us on this one!

- Agree on model outputs. Who needs what information when and in how much detail?

- Define the process for consistent information delivery. You can deploy plenty of processes; for example PAS 1192-2 and 3.

- Agree on data exchange formats and undertake interoperability checks. Will all the various software solutions work well together? Consider the use of open data formats such as Industry Foundation Classes (IFC).

- Define shared coordinates. Ensure that the whole team is working to the same agreed origin point.

- Determine data segregation and breaks in design to ensure that your model environment has a logical work breakdown structure and that each model will be of a manageable size and capable of being linked.

- Plan to ensure that you have adequate resources to create, analyze, validate, and exchange all this project data.

✔ **Pull your BIM project all together into a BIM execution plan (BEP).** Keep this document live and use it throughout the project to help communicate a consistent BIM strategy and modeling standards.

✔ **Establish your BIM capability and capacity.** Check that the proposed project team members have the right level of experience to deliver your information needs. What upskilling and training is required for both your team and perhaps that of your supply chain?

✔ **Manage object libraries.** Check to ensure that you're using libraries of objects and families that are appropriate to your project and that align with your levels of definition at the various project stages.

✔ **Pat yourself on the back.** Congratulations — you're now working on your project information model! Do so safely and ensure you follow cybersecurity standards.

✔ **Have regular team meetings to share and discuss the models.** Keep BIM at the heart of the meetings; make it the campfire you all sit around to discuss progress and project issues.

✔ **Agree on levels of detail and levels of information.** Check that the project team is building in manufacturer data such as warranty information as the project matures and key supply chain packages as procured.

✔ **Undertake regular project audits.** These ensure that the project team is following the BEP.

✔ **Gain knowledge from your projects.** Capture the good, bad, and ugly and use that knowledge to help refine your organizational BIM strategy.

Appendix C

Examining BIM and Education

*T*rying to apply BIM processes in industry has highlighted that a skills gap exists, and it's only going to get worse because of new technologies on the horizon. The best chance of creating an environment ready for collaboration is by engaging the next generation of construction professionals who are studying now. Students need to graduate with high levels of BIM awareness and that responsibility falls to colleges and universities. So if you're a teacher or lecturer, it's up to you.

In a nutshell, the sooner you can help put cooperation into practice, the more it becomes engrained in people. The more normal it becomes at college, the more likely people will carry it into the working world. The landscape is changing all the time, so BIM courses need to stay up to date and even lead the way through cutting-edge research. Education is something we're all really passionate about, and this appendix gives you some great ideas about how to teach BIM.

Getting Up to Speed with BIM

A lot of construction strategies and example projects are available that force the industry to realize that it can be more efficient, better managed, and more sustainable. BIM and integrated project delivery (IPD) requirements are integral to achieving these new efficiencies, but they're going to change a traditional industry in a big way. BIM isn't just about using the software; it's about practice management and data analysis. So education providers must supply the industry with graduates who are prepared for a new way of working.

In 2011, the UK government set out its intention to develop BIM learning outcomes, which provide you with a set of things you should be aiming to understand as a result of BIM education. The UK government will release the latest set of learning outcomes as part of the BIM Level 2 suite of documents.

BIM education and BIM training can be almost interchangeable terms in the public domain. Events promoted as BIM education are often directed at end users and construction professionals, not at students at all. Also, the industry often looks to academia to provide training and certification.

Keeping updated is incredibly important for most courses, but BIM does feel like a subject that's constantly shape shifting. As soon as you can put together a curriculum, another key document or process tool appears, continuing the evolution. Just like in industry, the project team is going to require quite a culture change. Instead of education providers planning everything on the curriculum to the nth degree and using it for years, a period of flux will occur in which you will need to review learning objectives and priorities more regularly.

Do you feel confident in teaching your students about BIM? We hope that this book gives you a good head start, but no shame exists in admitting that you may need further industry training to build your knowledge too, especially if your time in academia means you've been out of practice for a bit.

Don't forget that a lot of vendors provide educational versions of their software packages to academics. Often these contain full functionality, but outputs may be automatically watermarked to prevent commercial use.

Understanding Collaboration

You may have noticed how defensive people can get about their roles, titles, qualifications, and professional memberships, acting like cliques at a party or, worse, like armies in battle. For too long, education for construction has reinforced these silos, usually by funneling people into courses that lead to these discrete roles.

An academic situation allows people to experiment and try out things that they may not get away with in practice, and your BIM curriculum needs to provide the kind of environment where mistakes are allowed and actually form part of the learning process.

Perhaps you already include many group-work elements in your existing courses, because teams are (sometimes) more than the sum of their parts. BIM is all about collaboration, so teaching your class new processes should start with getting people working together, especially across disciplines and with groups they may not traditionally have encountered.

If you're part of a group of professors or lecturers teaching BIM and related subjects, you need to make sure that you're demonstrating collaboration and good communication to your students. If you can't bear to work with a certain professor or dean, what kind of an example is that?

Teaching BIM

One of the great benefits of teaching BIM processes and fostering collaborative environments is that these are transferable skills that your students can apply to many situations and whatever future careers they follow.

You need to integrate BIM into various courses, but not a lot of guidance exists on how you should achieve it. BIM can really disrupt what you've been used to.

A typical project may have 20 to 30 different roles. Making sure that your class understands what all the different players do and how they work together is important. The more you can introduce your students to multidisciplinary working, the better. It's the only way the industry will see the change it needs.

If you work with a large group of students, you may need to break the students into smaller groups. Doing so is a great opportunity to have them role-play at the various collaborative relationships of a typical project. Getting your students to take roles different to their studies and including clients, planners, building control, lawyers, and so on helps them to see the effects of their decisions on others. Perfect outputs are unrealistic if they represent only solo work, because that situation doesn't exist in reality.

Obviously, parts of BIM studies focus on the software and systems, but students gradually learn these skills over time, and you don't directly assess or test for them. You have to build any approach on the same foundation as BIM implementation: to encourage and embed information sharing and cooperative work.

Just choosing a BIM platform is daunting enough, and the expertise you need not just to be a proficient user but also to configure personalized settings within the software can require a very specific set of skills. Some vendors produce their own training and educational resources, but explaining things simply to others may take a lot of time and effort.

Highlighting Great BIM Teachers

Perhaps the best way to work out what requirements and skills are most important in education is to look at what others are doing. Here are some great organizations that deal with the same issues as you:

- ✔ **Pennsylvania State University** (http://bim.psu.edu): Penn State's Architecture and Engineering program is highly respected for its very collaborative approach and sustained investment in new technology and simulation spaces. They also produce BIM resources for industry.

- **Georgia Institute of Technology** (www.gatech.edu): Many BIM experts consider Georgia Tech to have one of the longest-running BIM and collaboration courses around. They represent the leading edge in process and technology.

- **Auburn University** (http://cadc.auburn.edu): Demonstrating how to collaborate across institutions and the globe, Auburn's BIM courses are jointly organized with international partners. The school's Master of Integrated Design and Construction led the way in college education for IPD.

- **CODEBIM** (http://codebim.com): This is a joint venture between three Australian universities. Jennifer McDonald at UTS is developing a framework for the implementation of collaboration techniques in BIM courses and publishing a wealth of content on the CodeBIM site.

- **UK BIM Academic Forum** (www.bimtaskgroup.org/bim-academic-forum-uk): This group is made up of representatives from more than 30 UK universities as a BIM promotion and collaboration group. You can see the group's progress in developing a standard academic framework for BIM knowledge sharing. Many members are also part of industry consultancy and exchange ventures like thinkBIM at Leeds Metropolitan and BIM Academy at Northumbria.

- **Class of Your Own (COYO;** http://designengineerconstruct.com**):** Alison Watson and COYO have advocated BIM to schools in the UK for years through the "Design, Engineer, Construct!" curriculum. Pupils as young as 10 are learning how to collaborate, use BIM software, and even write code. Some of the work the kids produce is really impressive and far beyond their years.

For a long time, industry has suggested that educators haven't done a great job of preparing students for the realities of practice. Perhaps the tracks of theory and industry have grown too far apart. You have a rare opportunity to equip your students to exploit a digital world.

So, everyone needs to quickly find the best way of building the skills and knowledge necessary to make construction more efficient and more innovative. In reality, doing so will only happen if industry and education work together. You know, if people start to collaborate. Sounds good, right?

Index

About the Authors

Stefan Mordue is a chartered architect, qualified construction project manager, Building Information Modeling (BIM) manager, visiting lecturer, and co-author of the literary classic *BIM for Construction Health and Safety*. He began his career with a number of small architectural practices before joining the NBS as a technical author, and he was instrumental in the development of the NBS National BIM Library and authored the NBS BIM Object Standard.

He was part of the team that produced a BIM Toolkit for the UK government as part of its BIM mandate, and more recently he has worked as a consultant to support organizations to implement NBS products and services. Stefan sits on a number of industry and technical standard committees, including the Architects' Council of Europe (ACE) BIM Working Group, and was a founding member of the CIC BIM2050 working group. A retired Ironman triathlete and native Londoner, he now lives in northern England with his wife and two children. You can follow on Twitter him at @StefanMordue.

Paul Swaddle is a chartered architect, urban futurist, and author. In his role as NBS Business Solutions Consultant, he's developed BIM frameworks to support the NBS products' ecosystem. He likes to find familiar analogies for complex subjects to make them easier to understand and has delivered hundreds of hours of training to construction professionals and students on specification and BIM.

He's passionate about continuous improvement in the built environment and the supply chain. He's a creator and collector of images and sounds, and he loves coffee, conversation, and American football. Above all, he's a hoarder of books and can always be found reading, writing, and illustrating. Follow him at @paul_swaddle.

David Philp is a Fellow of the Royal Institute of Chartered Surveyors (RICS), Chartered Institute of Building (CIOB), and the Chartered Institution of Civil Engineering Surveyors (ICES). He's also a chartered construction manager and RICS Certified BIM manager. Well known in the global construction sector, David's enthusiasm lies in highlighting the potential of new technologies and how the construction industry interacts with them to bring added value to customers and unlock new ways of digital working.

An early adopter of practical change and purposeful collaboration, he's director of BIM for Europe, the Middle East, Africa, and India for AECOM, the global, fully integrated services firm. He was seconded into the UK Cabinet Office in 2011 as head of BIM, a role he still fulfills today under the UK BIM Task Group. A founder of the CIC BIM 2050 Group, chair of the BIM4 communities, and professor at Glasgow University, he lives in Scotland. When he's not on a plane, he pursues his love of rodeo and Victorian poetry. Follow him at @ThePhilpster.

Dedication

From Stefan: To Becky, Isabelle, and Rosa, you keep my dark days away. I love you all more than you know.

From Paul: For my family, who taught me about collaboration, communication, and sharing long before I'd ever heard the term BIM. To Ruth and Ella, for all the time I've spent at my laptop and not with you.

From David: To Lori and Sophie — hi, girls; love you both! For all those who inspire change, challenge everything, and laugh at my jokes, this is for you guys.

Authors' Acknowledgments

This book is a great example of federation. Three of us wrote, and we've federated our individual content together in order to co-author the book. This has involved a lot of discussion and restructuring to see how everything fits together, and we all want to thank our co-authors for the opportunity to work together (and still be friends at the end). What you have in your hand (or on your screen) is federated information.

We all thank our colleagues and friends at AECOM, UK BIM Task Group, NBS, and RIBA Enterprises, and across the global construction industry, with a particular shout-out to the Twitter #UKBIMCrew for advice, debate, and inspiration throughout.

At John Wiley & Sons and their associates, we thank Annie, Michelle, Chad, and Charlie for filling our inboxes every single day with guidance, advice, encouragement, and near-constant pestering at deadlines!

Paul would also like to thank Josie Long and Ellie McDowall for the BBC podcast Short Cuts, which has most often been his soundtrack to writing this book. Thanks, Josie, for letting Paul use this quote, pinned to his desk since day one:

> *I'm absolutely an idealist and quite often I get accused of being unrealistic or naive with it. But for me, allowing yourself to believe in a better future, to step out of reality in that way, feels very powerful; like a small, daily act of resistance.*

> *Josie Long*

Publisher's Acknowledgments

Acquisitions Editor: Annie Knight

Project Manager: Michelle Hacker

Development Editor: Chad R. Sievers

Copy Editor: Charlie Wilson

Art Coordinator: Alicia B. South

Production Editor: Suresh Srinivasan

Cover Photos: ©Getty Images/teekid

Take Dummies with you everywhere you go!

Whether you're excited about e-books, want more from the web, must have your mobile apps, or swept up in social media, Dummies makes everything easier.

Visit Us

Like Us

Follow Us

Watch Us

Join Us

Pin Us

Circle Us

Shop Us

FOR DUMMIES®

A Wiley Brand

BUSINESS

978-1-118-73077-5

978-1-118-44349-1

978-1-119-97527-4

MUSIC

978-1-119-94276-4

978-0-470-97799-6

978-0-470-49644-2

DIGITAL PHOTOGRAPHY

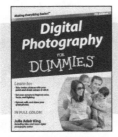
978-1-118-09203-3

978-0-470-76878-5

978-1-118-00472-2

Algebra I For Dummies
978-0-470-55964-2

Anatomy & Physiology For Dummies, 2nd Edition
978-0-470-92326-9

Asperger's Syndrome For Dummies
978-0-470-66087-4

Basic Maths For Dummies
978-1-119-97452-9

Body Language For Dummies, 2nd Edition
978-1-119-95351-7

Bookkeeping For Dummies, 3rd Edition
978-1-118-34689-1

British Sign Language For Dummies
978-0-470-69477-0

Cricket for Dummies, 2nd Edition
978-1-118-48032-8

Currency Trading For Dummies, 2nd Edition
978-1-118-01851-4

Cycling For Dummies
978-1-118-36435-2

Diabetes For Dummies, 3rd Edition
978-0-470-97711-8

eBay For Dummies, 3rd Edition
978-1-119-94122-4

Electronics For Dummies All-in-One For Dummies
978-1-118-58973-1

English Grammar For Dummies
978-0-470-05752-0

French For Dummies, 2nd Edition
978-1-118-00464-7

Guitar For Dummies, 3rd Edition
978-1-118-11554-1

IBS For Dummies
978-0-470-51737-6

Keeping Chickens For Dummies
978-1-119-99417-6

Knitting For Dummies, 3rd Edition
978-1-118-66151-2